Get the eBook FREE!

(PDF, ePub, Kindle, and liveBook all included)

We believe that once you buy a book from us, you should be able to read it in any format we have available. To get electronic versions of this book at no additional cost to you, purchase and then register this book at the Manning website.

Go to https://www.manning.com/freebook and follow the instructions to complete your pBook registration.

That's it!
Thanks from Manning!

The Jamstack Book

The Jamstack Book

BEYOND STATIC SITES WITH JAVASCRIPT, APIS, AND MARKUP

RAYMOND CAMDEN
BRIAN RINALDI
FOREWORD BY MATHIAS BIILMANN CHRISTENSEN

MANNING
SHELTER ISLAND

For online information and ordering of this and other Manning books, please visit www.manning.com. The publisher offers discounts on this book when ordered in quantity. For more information, please contact

Special Sales Department
Manning Publications Co.
20 Baldwin Road
PO Box 761
Shelter Island, NY 11964
Email: orders@manning.com

Manning Publications Co.
20 Baldwin Road
PO Box 761
Shelter Island, NY 11964

Development editor: Katie Sposato Johnson
Technical development editor: Louis Lazaris
Review editor: Mihaela Batinić
Production editor: Andy Marinkovich
Copy editor: Michele Mitchell
Proofreader: Keri Hales
Technical proofreader: David Cabrero
Typesetter: Gordan Salinovic
Cover designer: Marija Tudor

ISBN 9781617298882
Printed in the United States of America

To Lindy: You like to tell me to choose happy—and I absolutely did when I chose you. I love you. Also, Laser says Utini!

—Raymond

To my wife, Claudia, and my sons, Luke and Sam, who love me, challenge me, and give my life meaning. I love you!

—Brian

brief contents

contents

foreword

It's a great pleasure for me to write this foreword to Brian and Raymond's new book on the Jamstack. Both Brian and Raymond have been part of this dynamic movement that's changed the face of the modern web over the last 5 to 10 years.

I first met Brian at the start of 2015 when he was speaking about static site engines at the HTML 5 meetup in San Francisco. This was in the earliest days of Netlify, while the product was still in private beta, and before I had even coined the term *Jamstack*, at a time when just a few early adopters across the industry had started to believe that the web could be simpler, faster, safer, and better to develop with if we embraced the idea of decoupling the web UI from backend infrastructure and business logic.

It was meeting with and talking to these early adopters in different areas of our industry, like Brian, Raymond, and many others working on SaaS applications, headless CMSs, real-time web databases, interactive experiences on the web, and so on, that helped my cofounder and I build conviction that there was a broad, industry-wide change about to happen and that we needed a name for it and a nomenclature around it.

One night, in conversation with a friend, I came up with the term *Jamstack* and the rest—as they say—is history. We started circulating this term among the people we had connected with as well as the community that was already starting to form around Netlify at the time—Brian and Raymond among them—and the term started spreading.

Today the Jamstack ecosystem is at an interesting point of potential and tension. On the one hand, there's no doubt that the Jamstack architecture has changed the modern web for the better: we've seen a groundswell of platforms, frameworks, APIs, web databases, content, and commerce platforms emerge and grow up around the category, and the web is a second-to-none platform to develop for today. On the other hand, some of the initial principles around simplicity and of prebaking a ready-to-serve frontend are being challenged by different approaches to on-demand edge-based rendering layers and hybrid build tools, where developers sometimes have to navigate different rendering models on a page-by-page basis.

I've always known Brian and Raymond as curious minds always searching for simple and approachable tool chains that stay close to the fundamentals of the web. And this book is very much a practitioner's guide to the Jamstack, where you can take a tour with these two experienced Jamstack developers through a selection of the different build tools and site generators you'll encounter in today's Jamstack landscape, learn how to get productive, and make your own choices among different approaches for different projects.

The landscape of tools and frameworks will always be shifting and changing on the web, and no tool will ever be the right one for every problem you will encounter. Building a solid understanding of the strengths and tradeoffs that each tool brings, and developing a good instinct for how each tool feels, will help you better evaluate and navigate the rapidly evolving Jamstack ecosystem as existing frameworks change and innovate, and as new tool chains emerge.

—Mathias Biilmann Christensen
CEO and cofounder, Netlify

preface

Both Brian and I have been fortunate to have been in the web development business for many years. We've seen the good (evergreen browsers!), the bad (you'll get those new features in a year or two!), and the even worse (tables for layout totally make sense). As developers with a bit of experience (and gray hair), we both were excited about the introduction of the Jamstack (or static sites, as it was originally called). In many ways, it was a modern take on building the web that we fell in love with at the start of our careers—one based on simple files that we carefully handcrafted in text editors. But the Jamstack was also much more practical and sensible by making use of the best aspects of modern development frameworks and adopting modern best practices.

As proponents of the Jamstack, we both feel that there's no better time for developers to get involved. While the Jamstack has been around for a few years now, as a whole, it's still very new, and the tools and technologies around it are just now becoming mature and gaining mainstream adoption. We both think that the Jamstack is a compelling framework for building websites that's likely to be useful for most developers. That being said, there's also a strong need for a book that introduces readers to the Jamstack while also giving them multiple examples of what can be built with it.

As we're different people with different opinions, you'll see different tools discussed here, which will hopefully give you an appreciation for the breadth of options available to the aspiring Jamstack developer. We both agree that there's almost always a way to solve a problem using the Jamstack; the crucial bit is finding the solution that works best for you and your team. While going through the examples in this book, try to focus on the problems being solved, and if you aren't necessarily happy about the "how," be happy with the knowledge that there are always alternative solutions that can work much better for you.

—RAYMOND CAMDEN

acknowledgments

This book, like any good (and hard-won) project, was the work of many people. First, we would both like to thank our esteemed (and very patient) editor, Katie Johnson. Thank you for taking the time to explain the processes as well as putting up with our (totally valid!) excuses. We also appreciate the hard work of all the people at Manning who helped produce this book.

Raymond would like to thank Brian for agreeing to work with him, even though his instincts told him otherwise. The next book will be even more fun, right?

Brian would like to thank Raymond for pushing him into writing this book, and remains grateful even if he complained a lot during the process.

We'd both like to thank all the folks in the Jamstack community who work on the tools or write the articles that this book was built on. No technology can succeed without a great community behind it, and Jamstack definitely has that.

To all the reviewers: Alex Lucas, Amit Lamba, Anshuman Purohit, Baskar Rao Dandlamudi, C. Daniel Chase, Casey Burnett, Conor Redmond, David Cabrero, Fabrice Gouédard, Frans Oilinki, George Thomas, Jason Gretz, John McNew, Jonathan Cook, Mario Ruiz, Matej Strašek, NaveenKumar Namachivayam, Rodney Weis, Sachin Singhi, Satej Sahu, Scott Stroz, Sergio Arbeo, Sheik Uduman Ali M, Steve Albers, Theo Despoudis, Tristan V. Gomez, and Zoheb Ainapore, your suggestions helped make this a better book.

about this book

The Jamstack Book was written to help readers get an understanding of what working with the Jamstack really entails while also providing multiple real-world examples that apply that understanding. It begins by giving several different examples of building common website archetypes using popular static site generators. We then explore other parts of the Jamstack ecosystem that will be useful once developers start creating real-world projects using Jamstack tools.

Who should read this book

The Jamstack Book is for web developers who are looking to embrace, or at least consider, working with the Jamstack methodology. The reader should have a basic understanding of web fundamentals (HTML, JavaScript, and CSS) but need not be an expert in any particular aspect. This book will help cement the many benefits of using the Jamstack by providing multiple examples, such as blogs and documentation sites.

How this book is organized: A road map

The book has 10 individual chapters:

- Chapter 1 explains what exactly is meant by Jamstack and why developers should consider it.
- Chapter 2 introduces Eleventy and demonstrates a very simple brochure-ware site.
- Chapter 3 introduces Jekyll and walks you through building a blog.

- Chapter 4 makes use of the Hugo static site generator and explains how to build a documentation site.
- Chapter 5 demonstrates e-commerce with the Jamstack and uses Next.js.
- Chapter 6 explains how Jamstack sites can be moved into production.
- Chapter 7 demonstrates adding dynamic elements back into static web pages.
- Chapter 8 introduces serverless computing, with a focus on how it complements the Jamstack.
- Chapter 9 talks about how CMSs (content management systems) can be integrated with the Jamstack.
- Chapter 10 wraps things up with a look at how you can migrate to the Jamstack.

Developers can choose to read the book directly from beginning to end; however, chapters 2 through 5 serve as examples of different types of sites that can be built with the Jamstack and can be read in any order.

About the code

Installation instructions for the tools used in this book can be found as follows:

- Chapter 2 covers the installation of Eleventy. The latest installation instructions can be found at https://www.11ty.dev/docs/getting-started/.
- Chapter 3 covers the installation of Jekyll. The latest installation instructions can be found at https://jekyllrb.com/docs/installation/.
- Chapter 4 covers the installation of Hugo. The latest installation instructions can be found at https://gohugo.io/getting-started/installing/.
- Chapter 5 covers the installation of Next.js. The latest installation instructions can be found at https://nextjs.org/docs/getting-started.

This book contains many examples of source code, both in numbered listings and in line with normal text. In both cases, source code is formatted in a `fixed-width font like this` to separate it from ordinary text. Sometimes code is also **`in bold`** to highlight code that has changed from previous steps in the chapter, such as when a new feature adds to an existing line of code.

In many cases, the original source code has been reformatted; we've added line breaks and reworked indentation to accommodate the available page space in the book. In rare cases, even this was not enough, and listings include line-continuation markers (➥). Additionally, comments in the source code have often been removed from the listings when the code is described in the text. Code annotations accompany many of the listings, highlighting important concepts.

You can get executable snippets of code from the liveBook (online) version of this book at https://livebook.manning.com/book/the-jamstack-book. The complete code for this book can be downloaded from the GitHub repository at https://github.com/cfjedimaster/the-jamstack-book.

liveBook discussion forum

Purchase of *The Jamstack Book* includes free access to liveBook, Manning's online reading platform. Using liveBook's exclusive discussion features, you can attach comments to the book globally or to specific sections or paragraphs. It's a snap to make notes for yourself, ask and answer technical questions, and receive help from the author and other users. To access the forum, go to https://livebook.manning.com/book/the-jamstack-book/discussion. You can also learn more about Manning's forums and the rules of conduct at https://livebook.manning.com/discussion.

Manning's commitment to our readers is to provide a venue where a meaningful dialogue between individual readers and between readers and the author can take place. It is not a commitment to any specific amount of participation on the part of the author, whose contribution to the forum remains voluntary (and unpaid). We suggest you try asking the authors some challenging questions lest their interest stray! The forum and archives of previous discussions will be accessible from the publisher's website as long as the book is in print.

Other online resources

There are multiple places where folks can learn more about the Jamstack and keep up on the latest changes. Here's a list of resources to consider:

- Jamstack.org is a great high-level website about Jamstack, with many links to other resources. You can join its Discord channel as well.
- The New Dynamic (https://www.tnd.dev/) is another great "meta" resource with its own Slack.
- JAMstacked (https://jamstack.email/) is a weekly newsletter Brian curates with the latest news on Jamstack blogs, events, and more.

about the authors

RAYMOND CAMDEN is a senior developer evangelist for Adobe. He works on the Document Services APIs to build powerful (and typically cat-related) PDF demos. He is the author of multiple books on web development and has been actively blogging and presenting for almost 20 years. Raymond can be reached at his blog (www.raymondcamden.com), @raymondcamden on Twitter, or via email at raymondcamden@gmail.com. He's married with eight kids (yes, you read that right) and multiple furry creatures.

BRIAN RINALDI is a developer experience engineer at LaunchDarkly with over 20 years of experience as a developer for the web. Brian is actively involved in the community running developer meetups via CFE.dev and Orlando Devs. He's also the editor of the JAM-stacked newsletter, a biweekly Jamstack focused newsletter.

about the cover illustration

The figure on the cover of *The Jamstack Book* is “Femme des Environs de Rome,” or “Woman from the Surroundings of Rome,” taken from a collection by Jacques Grasset de Saint-Sauveur, published in France in 1797. Each illustration is finely drawn and colored by hand.

In those days, it was easy to identify where people lived and what their trade or station in life was just by their dress. Manning celebrates the inventiveness and initiative of today’s computer business with book covers based on the rich diversity of regional culture centuries ago, brought back to life by pictures from collections such as this one.

1 Why Jamstack?

This chapter covers

- Defining Jamstack as an architecture for web applications rather than a prescriptive stack of technologies
- How Jamstack formed in response to dynamic web page development that had become cumbersome, slow, and insecure
- Benefits of Jamstack, including page speed, security, and cost
- Exploring well-known websites that are built with the Jamstack

As Jamstack has gained popularity in recent years, a common criticism lobbed at it is that it is just a marketing term. The truth is that they are right. As we'll explore, *Jamstack* was a term invented to "rebrand" an architecture many developers were already using to build sites because the existing terminology had become misleading. While calling it marketing may be a fair critique, Jamstack is still a way of building sites that has been gaining rapid adoption by web developers.

1.1 What is the Jamstack?

The Jamstack is not a simple thing to define. There is no Jamstack installer. There's no predefined set of tools you should install that comprise the Jamstack. There's not even a specific language associated with developing Jamstack apps. (Yes, JavaScript plays a central role, but any number of languages may also be involved, including Ruby, Go, Python, or others.) Ultimately, there are countless combinations of tools and languages that can be combined to create a site that could legitimately be called Jamstack.

What Jamstack is instead is more of an architectural pattern or methodology for creating sites. While there is a lot of ongoing debate about this, these are the key elements as we define them:

- *A Jamstack site is primarily built on static assets.* Jamstack sites are always deployed as static files. This means that they are not dynamically generated by an application server when the user requests a page; instead, the site files are generated at build time. For a Jamstack site, every user who requests a specific page in their browser will get the same static asset returned. However, this does not mean the content is static. In fact, modern Jamstack sites offer an array of rendering options for the content of pages, including fully static and server-side rendering.
- *A Jamstack site is built using a static site generator.* The static assets in a Jamstack site are generated using a static site generator (SSG). At a very basic level, a SSG is a tool that takes templates and combines them with content. Content can be stored in files as Markdown, YAML, or JSON files or be pulled from APIs. The content and template combine to dynamically generate the site's HTML, CSS, and JavaScript assets. This is similar to the process a dynamic web server like PHP might go through on each user request, but, instead, the majority of this process happens at build time before the site is even deployed.
- *A Jamstack site leverages APIs.* What differentiates a Jamstack site from a simple static site is that, although it is comprised of static assets, it can be very dynamic. The first key ingredient to creating this dynamic functionality is the use of APIs. These APIs can be called by the browser client at run time or even called by the static site generator at build time.
- *A Jamstack site uses JavaScript for dynamic functionality.* The second key ingredient to making a Jamstack site dynamic is its ability to call APIs and other services asynchronously on the client via JavaScript. JavaScript is what allows the static assets to change dynamically via document object model (DOM) manipulation. Client-side JavaScript powers things like user logins or shopping carts.

Clearly there's a lot of flexibility in this definition, which, in my opinion, is part of Jamstack's appeal. There is almost certainly a combination of Jamstack tools and services that meet the needs of your project and your language, tooling, and deployment preferences.

That flexibility has a cost, though. There isn't a single way to teach someone Jamstack, and the multitude of options can make the learning curve for newcomers a bit steep. Also, there is arguably additional complexity in creating a site that leverages a variety of APIs and services while also dynamically updating content on the client using JavaScript.

So why choose Jamstack? The Jamstack evolved in part to address the problems of a dynamic web that many felt had become slow, costly, and insecure. To better understand the need for the Jamstack, we need to understand how and why it evolved.

1.2 A brief history of Jamstack

By learning why the term *Jamstack* was created in the first place, we can gain a better understanding about what it is and why it has been rapidly gaining popularity. This is especially true because while Jamstack is a modern architecture that leverages many of the latest trends in technology, in other ways it harkens back to the way we built pages when the web was just invented.

The earliest web pages were just simple HTML deployed to a web server. For example, the first website, as shown in figure 1.1, was just a basic static site. Every person who visited the site received the same assets.

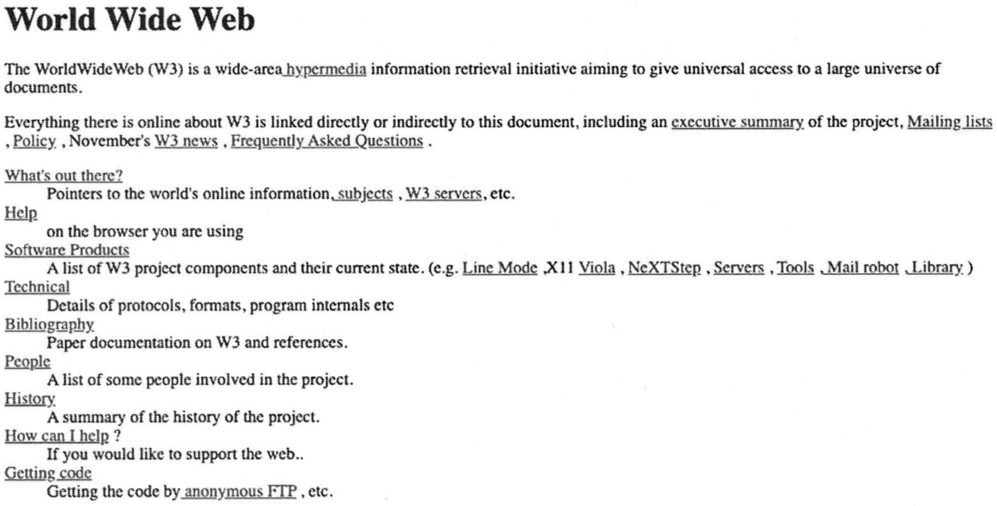

World Wide Web

The WorldWideWeb (W3) is a wide-area hypermedia information retrieval initiative aiming to give universal access to a large universe of documents.

Everything there is online about W3 is linked directly or indirectly to this document, including an executive summary of the project, Mailing lists , Policy , November's W3 news , Frequently Asked Questions .

What's out there?
    Pointers to the world's online information, subjects , W3 servers, etc.
Help
    on the browser you are using
Software Products
    A list of W3 project components and their current state. (e.g. Line Mode ,X11 Viola , NeXTStep , Servers , Tools , Mail robot , Library)
Technical
    Details of protocols, formats, program internals etc
Bibliography
    Paper documentation on W3 and references.
People
    A list of some people involved in the project.
History
    A summary of the history of the project.
How can I help ?
    If you would like to support the web..
Getting code
    Getting the code by anonymous FTP , etc.

Figure 1.1 The first web site was a static site. It is still available at http://info.cern.ch/hypertext/WWW/TheProject.html.

As the needs of the web evolved, so did the technologies that underpinned it. Web application servers and server-side scripting languages such as PHP and Ruby allowed sites to dynamically generate content. This allowed every user to be served custom assets that were dynamically rendered on the server before they were sent to the individual's browser. Today this is commonly referred to as *server-side rendering* (SSR).

Let's look at how a typical server-side rendered web application worked circa 2008 (why 2008? I'll explain in a moment):

- The user would request a page from the browser.
- The browser would hit the web application server, which would load the requested page built using some form of scripting language.
- The scripting language would make calls to the database for things like user information, product information, and/or content.
- The data and script would be combined to generate the HTML that was then sent to the user.

This process would be repeated on every page request. It allowed for highly personalized and dynamic content to be served from a single script file, but it came with costs:

- *Performance*—Each piece of this process entailed small performance costs, from the application server processing the request to the database processing queries, to generating the final HTML. Since this process repeated on each page request for every user, the costs could add up quickly and could be compounded when a web application server or database was under heavy load.
- *Security*—By nature these applications left a wide surface area open to potential attack. These could include things like vulnerabilities in the web application server to the scripting language or framework. The database could also be left open to direct attack through methods like SQL injection.
- *Scaling*—Since every request for every user required a unique response, these applications could become expensive and complex to scale as usage grew. Typically, servers were maintained in-house, so scaling meant new hardware, which meant that the application couldn't quickly or easily scale with need.

All these problems were solvable with the right resources, of course, but, at the time, so much of the web had become dependent on SSR that just viewing text content on a simple blog relied on an entire server-rendered architecture using tools like WordPress, which was growing rapidly in adoption.

1.2.1 The rise of static site generators

The year 2008 was pivotal for what would eventually become known as the Jamstack because it was the year that Tom Preston-Werner released Jekyll. Sure, there were static site generators before Jekyll, but Jekyll was based on the principles that have helped drive most modern static site generators since:

> *First, all my writing would be stored in a Git repository. This would ensure that I could try out different ideas and explore a variety of posts all from the comfort of my preferred editor and the command line. I'd be able to publish a post via a simple deploy script or post-commit hook. Complexity would be kept to an absolute minimum, so a static site would be preferable to a dynamic site that required ongoing maintenance.*
>
> —Tom Preston-Werner
> ("Blogging Like a Hacker," http://mng.bz/ExJO)

Jekyll gained widespread adoption, particularly as an alternative to WordPress for blogging. This adoption was driven in part by GitHub, of which Preston-Werner was a cofounder and CEO, when GitHub Pages added Jekyll support in 2008.

GitHub Pages support also introduced a new continuous development workflow that has since become ubiquitous in the modern Jamstack. Instead of running a local build and pushing the generated HTML, CSS, and JavaScript via FTP, developers could simply check in their changes to the GitHub repo, and the Jekyll build would be run and deployed for them.

Jekyll was followed by a long—and I do mean *long*—list of static site generators that followed similar principles. The most comprehensive list of static site generators (https://staticsitegenerators.net/) lists 460 as of this writing, with one built using almost every programming language available (even Swift, a language intended for developing iOS native apps, has a static site generator). These included some popular options still in widespread use today such as Middleman in 2009 (written in Ruby like Jekyll), Pelican in 2010 (written in Python), and Hugo in 2013 (written in Go).

1.2.2 From static sites to JAMstack

Back in 2016, Netlify was already a fast-growing startup focused on providing continuous deployment for developers using tools like static site generators, but the term *static sites* had become problematic. An ever-growing list of tools and services were enabling dynamic capabilities on sites built with static site generators. Static sites, it turned out, could be far from static in reality.

To solve this problem, Netlify's Matt Biilmann came up with a new term: *JAMstack*. This came with a new site, jamstack.org, that included his manifesto defining the new term. The original version he posted is shown in figure 1.2.

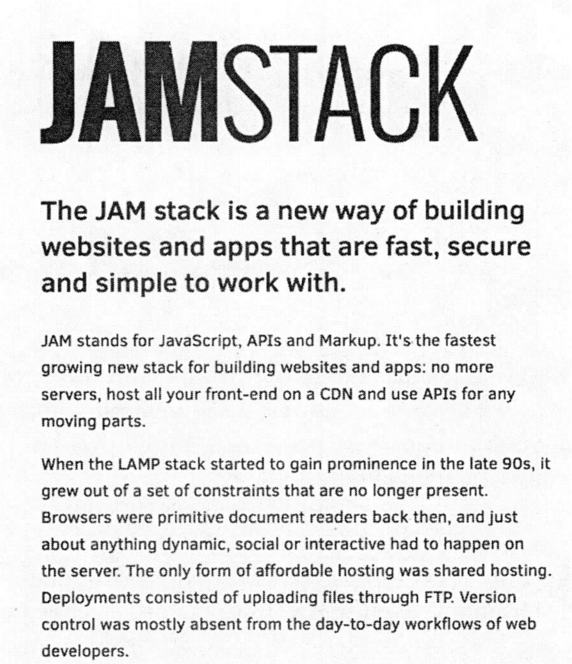

JAMSTACK

The JAM stack is a new way of building websites and apps that are fast, secure and simple to work with.

JAM stands for JavaScript, APIs and Markup. It's the fastest growing new stack for building websites and apps: no more servers, host all your front-end on a CDN and use APIs for any moving parts.

When the LAMP stack started to gain prominence in the late 90s, it grew out of a set of constraints that are no longer present. Browsers were primitive document readers back then, and just about anything dynamic, social or interactive had to happen on the server. The only form of affordable hosting was shared hosting. Deployments consisted of uploading files through FTP. Version control was mostly absent from the day-to-day workflows of web developers.

Figure 1.2 The original jamstack.org site launched in 2016.

The JAM in JAMstack stood for the following:

- *JavaScript*—This is the key to much of the dynamic capabilities of these sites. JavaScript enabled things like the asynchronous loading of content from APIs and the dynamic updating of the HTML on the client.
- *APIs*—These could be anything from a preexisting API provided by a third party to a cloud function to perform custom business logic. APIs provided these sites with the data and business logic that they needed.
- *Markup*—This comprised everything from the Markdown and YAML/TOML that contained the site content to the templating language (Liquid, Handlebars, etc.) that is used by the static site generator to generate the HTML pages. The static site generator is critical to this aspect, even if its presence is somewhat obfuscated in the JAM acronym—perhaps by design.

With Netlify's influence in the ecosystem, along with the cooperation of other companies in the space, the hope was that the term would help redefine how these tools were viewed in the web developer community. This was given a big assist with the launch of the JAMstack Conference in May of 2017 (another initiative led by Netlify), which has since spawned numerous follow-ups around the world.

JAMstack was—and still is—criticized for being simply a marketing term. As we have seen, this has some truth to it, but the years since the term was introduced have seen enormous growth in adoption and a whole ecosystem of companies that began to target JAMstack developers.

1.2.3 From JAMstack to Jamstack

There's one final, seemingly small but ultimately important, change to the terminology to discuss. In early 2020, the team at Netlify who manage the jamstack.org site opened a discussion to change the way the term was written, from JAMstack to Jamstack. Many community members chimed in, and the decision was made to make the change. As of this writing, most companies and organizations have followed suit, but the usage isn't yet uniformly adopted.

It's worth understanding the reasoning behind this change. The JAM acronym was showing signs of becoming the source of some confusion. First, JavaScript, APIs, and Markup seem like things that could describe almost any site being built for the web; the acronym wasn't making the differentiation clear. Second, leading with JavaScript seemed to create the impression that Jamstack was synonymous with JavaScript frameworks and exclusively JavaScript framework–based static site generator tools. Finally, JAMstack is not really a "stack" in that there is no preset group of tools as in LAMP. In reality, it's more of an architecture or even a methodology.

The hope was that changing the capitalization of the term would de-emphasize both the acronym and the "stack" and extend the life of the term in much the same way people still use Ajax instead of AJAX, long after XML largely dropped out of the equation. From here on out in this book, we'll stick with the Jamstack capitalization of the word.

1.3 The benefits of Jamstack architecture

Now that we understand how the Jamstack evolved from the early days of simple static sites to a modern architecture for building complex sites, let's answer the question, "Why should you choose Jamstack?" Here are some of the key benefits.

1.3.1 Performance

There are three important aspects to the performance of Jamstack applications to understand:

- *Static assets load faster than dynamic ones.* There is no processing that needs to occur to turn dynamic templates into HTML, CSS, and JavaScript, and no database calls being made at run time. All of these assets are *pre-rendered,* to use a term common in the Jamstack community that means the majority of the page rendering occurs at build time rather than run time.
- *Jamstack sites are served "from the edge."* Since the assets are static, they can be served from CDNs, meaning that each end user is served the site assets from the server closest to them. This is combined with instant cache invalidation when a new version is released so that users always get the most up-to-date version of the site from the CDN.
- *Jamstack sites scale by default.* There's no need to create additional servers to accommodate a heavy traffic load when your site is being served from a CDN. Plus, Jamstack sites rely on services like cloud functions, which are built to scale, for dynamic processing and functionality.

1.3.2 Security

Security is a tough topic to make broad assertions about, as there is no such thing as a completely secure option. A properly patched and maintained WordPress site can be secure, but the reality is that recent data shows that "73.2% of the most popular WordPress installations are vulnerable" (http://mng.bz/7WRg). As figure 1.3 shows, thousands of the top websites are susceptible to known vulnerabilities.

These issues are not WordPress-specific. Traditional sites have a lot of moving parts that need maintenance and patching regardless of the content management and application framework. For example, for a WordPress or Drupal site these parts would include things like the PHP web server and the MySQL database. For a Django site, it might be Python and PostgreSQL.

In contrast, Jamstack sites benefit from a greatly reduced surface area for attacks. There is

- No web server to compromise
- No web application server or web application framework with potentially unpatched security flaws to exploit
- No database to gain access to
- No central source of truth to hack because the site is served from multiple CDNs

WordPress Version	No. of Installations	No. of Known Vulnerabilities
3.6	13,034	5
3.6.1 (latest)	7,814	0
3.5.1	6,859	8
3.5.2	4,031	0
3.4.2	2,204	12
3.5	1,655	10
3.3.1	820	24
3.2.1	820	10
3.3.2	732	14
3.4	295	15
Total (Excl 3.6.1)	**30,823**	

Figure 1.3 WordPress sites can be secure, but data from WP WhiteSecurity, which shows the version WordPress installs in the Alexa top sites, shows many remain vulnerable due to not being updated. However, Jamstack sites do not require these sorts of updates as the assets are static (http://mng.bz/7WRg).

Yes, Jamstack sites can depend on third-party services, which can be open to attack. However, this also gives Jamstack sites the ability to take advantage of the domain expertise of these services. For instance, rather than implementing a custom authentication, they can take advantage of services like Auth0 or Netlify Identity that specialize in authentication and implement industry best practices. Plus, as software as a service (SaaS), there is no need for the developer to worry about patching.

1.3.3 Cost

Okay, let's get to my personal favorite benefit of Jamstack: it can greatly reduce costs to the point of even being frequently free. Since there is no need for web application servers and database servers, the costs of hosting the Jamstack are generally negligible to nonexistent. Continuous deployment services like Netlify, Vercel, and Render all have generous free plans that can accommodate the needs of many sites and pricing and generally scale based on usage or additional features. Some services, like GitHub Pages, offer continuous deployment and hosting for free (with limitations, of course).

The same is true of many of the third-party services that are popular in Jamstack sites: commercial offerings have generous free tiers, and entirely free or open source options exist, though typically with some limitations. For example, services like Algolia for search or Sanity for content management offer free tiers that can make them workable as cost-free or low-cost options for many sites. Meanwhile, tools like Lunr for search and Netlify CMS for content management provide free and open source alternatives.

Services and hosting are just two of the places where Jamstack can offer potential savings. Many Jamstack case studies cite reduced development and maintenance costs. For example, a recent Netlify white paper cited Loblaw Digital lead time reductions for a single campaign being reduced to "a month instead of the typical year, representing a 10× reduction in time to market, [and] $38,000 monthly cost savings" (https://www.netlify.com/whitepaper/).

1.4 When Jamstack may not be the right choice

Recently, it has become far more difficult to make a clear-cut recommendation about when to use and when not to use Jamstack. The improved capabilities of Jamstack tools and services make almost any type of site possible using a Jamstack approach. In fact, tools such as Next.js, Nuxt, and Gatsby now make it possible to create a hybrid solution that gives developers the option to make some routes static and others server-side rendered. But there are times when a Jamstack approach may not make sense:

- *An application that relies heavily on user-generated content*—It is entirely possible to build user-generated content as static assets or pull it via an API. There are examples of Jamstack sites adding things like user-generated comments or posts that are written to files that trigger a rebuild or are pulled dynamically from an API. This can make sense in cases where the user-generated content is periodic and supplemental, but for sites primarily focused on user-generated content, a Jamstack solution may prove to be overly complex and difficult to implement.
- *An application where content is continually updated*—Similar to a site with user-generated content, a site with constantly updated content (e.g., a real-time news site) may not be ideal for a Jamstack approach. Yes, this content could be updated live via client-side API calls or via SSR, but this can be difficult to properly implement and may negate some of the overall performance benefits of the Jamstack.
- *A dashboard that relies heavily on server-side processing*—Some dashboard applications make perfect sense as Jamstack. In many cases, these dashboards call APIs to populate charts and data tables that make sense to process on the client side. However, in other cases this may unnecessarily put too heavy a load on the client and not be an optimal solution.

As you can see, the line for what can be a Jamstack application is blurry. Every one of these examples can be built as a Jamstack application. I'd think about it in terms of how complex the Jamstack solution would be over a traditional server-side solution and how much am I offloading work to client-side API calls or SSR versus content that is generated as static assets. Don't over-architect a site simply to make it fit into Jamstack paradigm, as you may find the solution brittle and difficult to maintain. Don't offload the majority of your content display to client-side API calls, as this can negate some of the primary benefits of Jamstack's static, CDN-based approach. These are guidelines, not hard-and-fast rules, but they can help you evaluate whether the benefits of Jamstack outweigh the drawbacks for your project. For most projects, I am certain they will, but in some cases they may not.

1.5 Popular sites built with the Jamstack

It can be difficult to tell if a project was built with the Jamstack. This is because

- The Jamstack site looks no different than its dynamic alternative. In fact, the idea is that you will not compromise on design or functionality by choosing the Jamstack.
- There are countless ways to build a Jamstack web app. For example, there are currently 460 static site generators listed on staticsitegenerators.net. Headlesscms .org, which lists headless CMS systems (content management systems typically associated with Jamstack sites), currently lists 74 options. Deployment options for Jamstack sites include Netlify, Vercel, Render, AWS, Azure, and more. A Jamstack site might combine any number of these tools and services or others.
- There are often no obvious indicators in the site's code that signify it was built with the Jamstack. Some popular tools like Gatsby can be detected by the JavaScript they load; others like Hugo and Jekyll can be detected by their meta generator tags, but still others like Eleventy cannot be detected at all.

Nonetheless, examples of successful sites can often be useful to help make the case for adopting the Jamstack. Here are just a few examples of sites you may recognize that are built with the Jamstack as of this writing.

1.5.1 Smashing Magazine

Smashing Magazine is a popular site for designers and developers that moved from WordPress to a Jamstack site built with the Hugo static site generator and Netlify CMS, a popular open source content management solution specifically for Jamstack sites (see figure 1.4). (You can read more about their transition here: http://mng.bz/NxEX.)

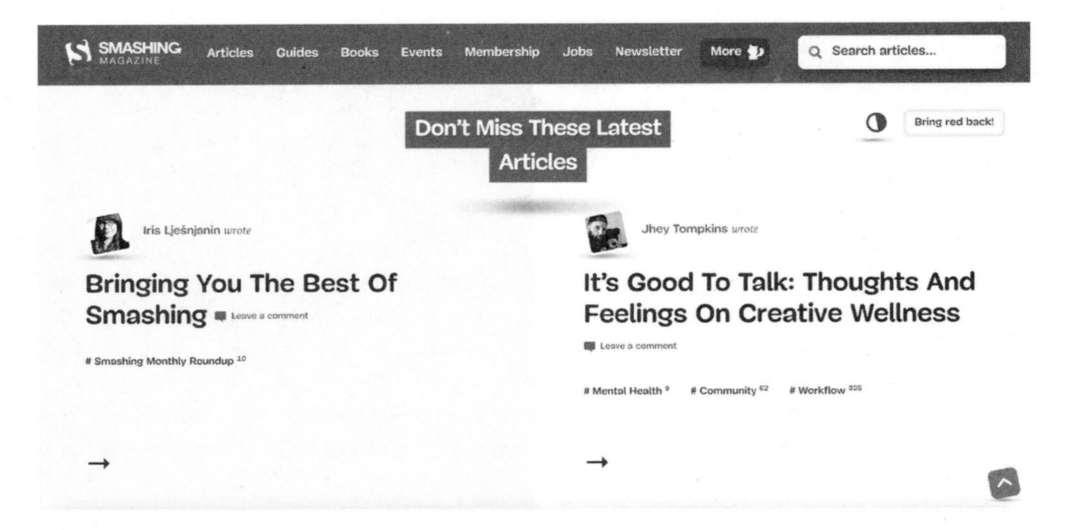

Figure 1.4 ***Smashing Magazine*** **relaunched their site using the Hugo static site generator and Netlify CMS for content management.**

1.5.2 *Nike*

Nike's site, including e-commerce, is built with Gatsby, a very popular React-based static site generator (see figure 1.5).

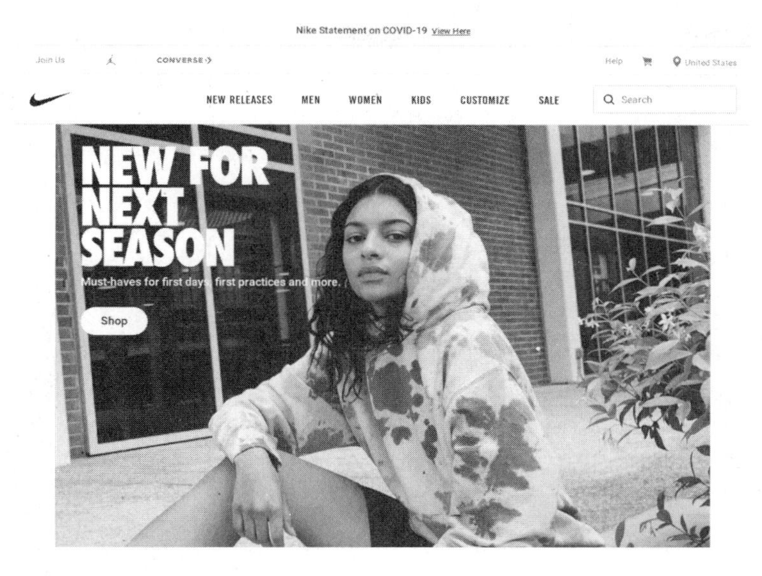

Figure 1.5 Nike.com built their site with Gatsby, a popular React-based static site generator.

1.5.3 *Impossible Foods*

A popular meatless food brand, Impossible Foods, also built their website with Gatsby (figure 1.6).

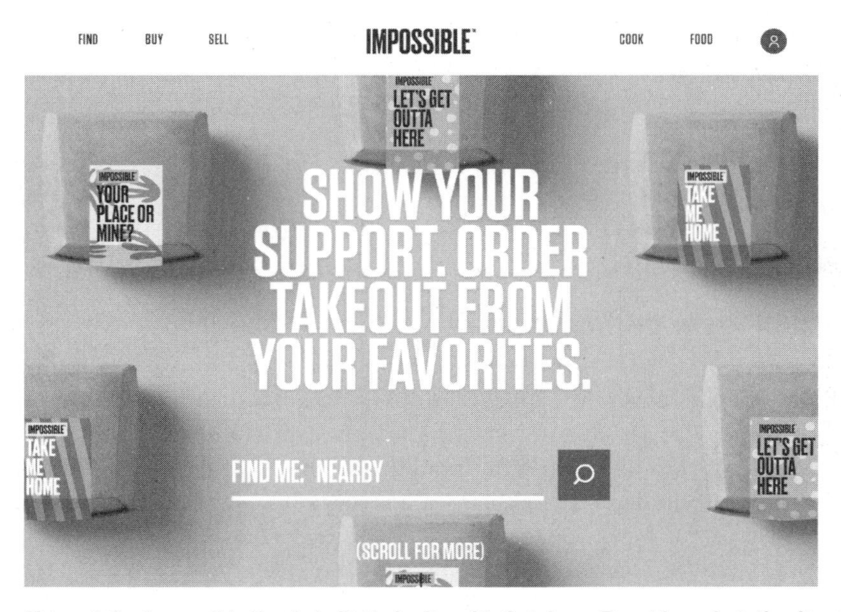

Figure 1.6 Impossible Foods built their site with Gatsby, a React-based static site generator.

1.5.4 Restaurant Brands International (RBI)

Restaurant Brands International (RBI), the company behind popular restaurants like Popeyes, Burger King, and Tim Hortons, moved their digital services to Jamstack in 2018. They spoke about it at Jamstack Conference in 2019. This allowed them to create a single platform for all their sites that shared components, content, and data across the various brand properties. In addition, the same content APIs that drive the site content on their Jamstack sites can be reused on the mobile apps (see figure 1.7).

Figure 1.7 Popeyes.com is among a number of web properties owned by RBI that were rebuilt using Jamstack technologies.

1.5.5 Digital.gov

Digital.gov is a government site that provides guidance, training, and community support to people who provide digital services in the US government. It was built with Hugo, a popular Go-based static site generator (figure 1.8). (You can read more about their case study here: https://gohugo.io/showcase/digitalgov/.)

1.6 What you'll learn in this book

In this chapter, we've seen why and how the Jamstack evolved as a solution to the issues that were plaguing web application development—starting as simple static site generators and evolving to the more complete application architecture that it is today. We now also understand some of the benefits of choosing to use the Jamstack and know examples of popular sites that use the Jamstack. But *how* can you use it?

The nature of the Jamstack means that there is no easy answer to that question. There are a myriad of combinations of tools and services you can use to create a site

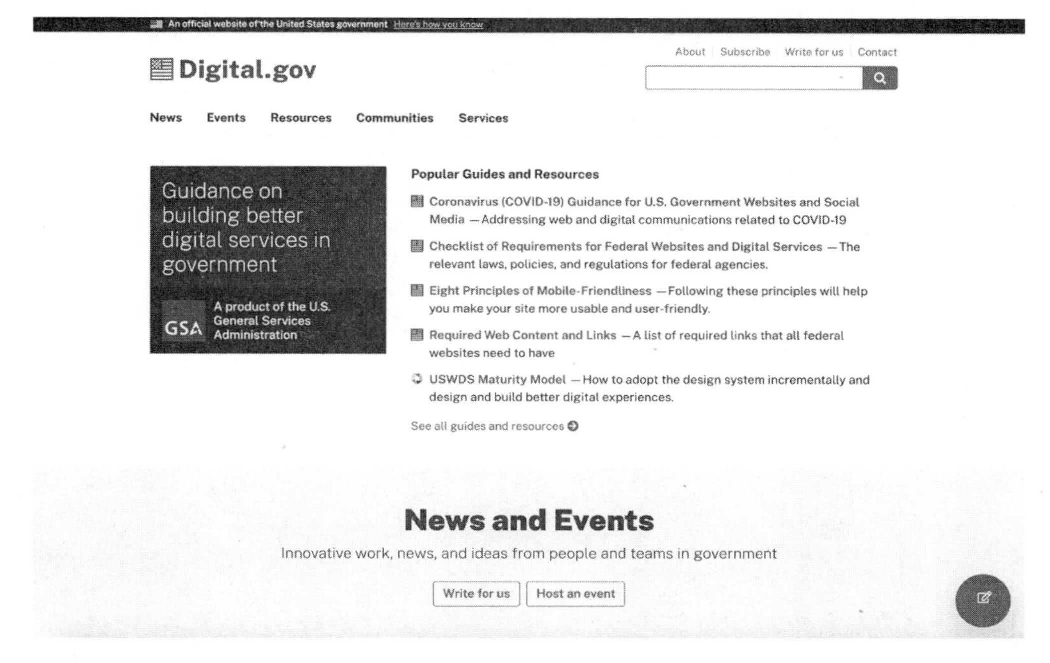

Figure 1.8 Digital.gov

that would qualify as Jamstack. This is why many books and tutorials about Jamstack tend to gravitate toward a single solution. In this book, we aim to take a different approach and offer you the opportunity to explore and use a wide variety of Jamstack tools and services to address real use cases. The goal isn't to make you an expert in each tool or service, but to offer you the experience you need to be able to evaluate what solutions are best suited for your project's needs and your personal preferences. Let's get coding!

Summary

- Jamstack is an architecture for building sites that combine markup, APIs, and templates to output static HTML, CSS, and JavaScript assets using a static site generator.
- Jamstack sites leverage a combination of APIs and JavaScript to achieve dynamic functionality equivalent to their dynamically generated counterparts.
- While the Jamstack term was created as a rebranding of "static sites," the concept evolved as a response to speed, cost, and security concerns around dynamic websites.
- The benefits of adopting Jamstack include performance (due to the use of static assets served from CDNs), security (due to the lack of a web server, web

application server, web application framework, and database server to compromise), and cost (due to the low cost of hosting static assets and the incremental costs of things like serverless computing).

- While Jamstack sites can be difficult to identify, many well-known companies have adopted the Jamstack architecture for website development.

2 Building a basic Jamstack site

This chapter covers

- Using the Jamstack to build a simple site
- How to install and make use of Eleventy
- Creating a coffee shop website using Eleventy

The first site we're going to build in the Jamstack will be what has previously been referred to as a *brochureware* website. In the past, this was used to describe websites that were nothing more than digital versions of a marketing brochure. While the web is an incredibly powerful platform and enables powerful collaboration, sometimes simple is all you need. There are multiple examples of where a one- or two-page site is more than satisfactory:

- A restaurant website that just displays opening hours, the address, and a menu (but not in a PDF please!)
- A temporary website for a service that is coming soon, perhaps with a simple form for people to sign up to be notified on release
- A landing page for a mobile app, with only links to the respective app stores

Each of these examples could easily be handled by a website with one to two pages. In fact, a static site generator may even seem like overkill for such a small project. But as many developers know, projects tend to grow over time. By starting off with a static site generator, you'll be much better prepared to adapt the website and add additional pages and features over time. For our first static site generator, we will use the incredibly flexible Eleventy.

2.1 Working with Eleventy

Eleventy (figure 2.1) is a simple but powerful static site generator that is incredibly flexible. Eleventy can be used to build just about anything you can imagine and isn't necessarily tailored to one kind of site versus another. This flexibility makes it an excellent first choice for developers learning the Jamstack. It doesn't have any preconceived notions about what kind of sites you will build and pretty much allows you to develop anything you need. Its wealth of templating languages also means you can use one that best matches your style of writing. It also relies on npm for installation, which is a very common tool among web developers, making the setup easier as well. This is why I've chosen Eleventy for the first static site generator covered in the book and also as the engine behind my personal blog, a site I've run for almost 20 years with over 6,000 blog posts. You can find Eleventy's home page at https://www.11ty.dev/.

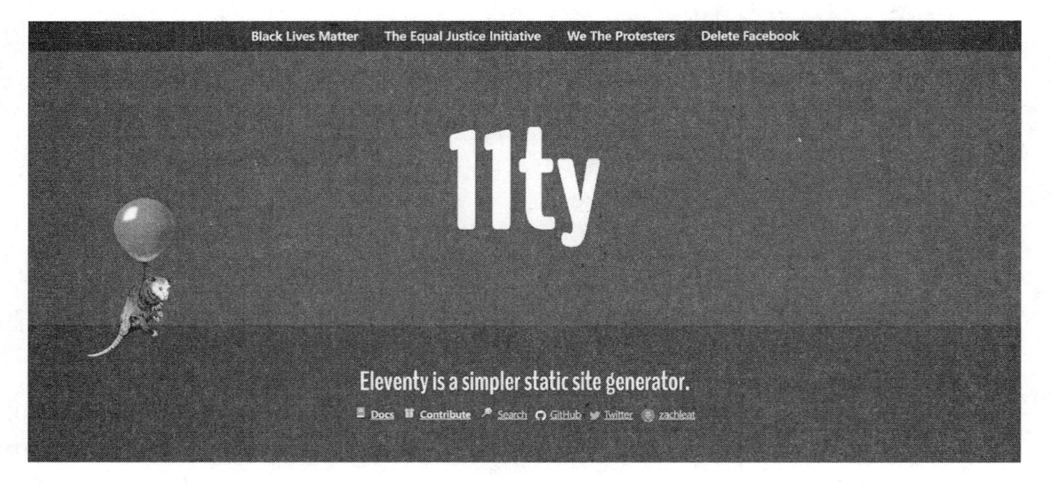

Figure 2.1 The Eleventy website

To begin working with Eleventy, you need to install the CLI tool via npm. *npm* stands for "Node package manager" and is a common tool to install utilities. You don't have to be a Node developer to use npm but installing Node (via nodejs.org) is the simplest way to get the npm installed. Assuming you have npm available in your terminal, you can install Eleventy's CLI like so:

```
npm install -g @11ty/eleventy
```

You can then confirm it worked by typing `eleventy --help` in your terminal, as shown in figure 2.2.

```
ray@Mandalore:~$ eleventy --help
usage: eleventy
       eleventy --input=. --output=./_site
       eleventy --serve

Arguments:
     --version
     --input=.
       Input template files (default: `.`)
     --output=_site
       Write HTML output to this folder (default: `_site`)
     --serve
       Run web server on --port (default 8080) and watch them too
     --watch
       Wait for files to change and automatically rewrite (no web server)
     --formats=liquid,md
       Whitelist only certain template types (default: `*`)
     --quiet
       Don't print all written files (off by default)
     --config=filename.js
       Override the eleventy config file path (default: `.eleventy.js`)
     --pathprefix='/'
       Change all url template filters to use this subdirectory.
     --dryrun
       Don't write any files. Useful with `DEBUG=Eleventy* npx eleventy`
     --help
ray@Mandalore:~$
```

Figure 2.2 Help output from the Eleventy CLI

The Eleventy CLI boils down to two basic features. It can either take a directory and build the source code into a static site or set up a web server that does that same thing but also allows for *hot reloading* (i.e., reload the browser after a file change is detected). Typically you would use the web server feature while building your site and perform builds when you just want the static file output of your site.

Eleventy lets you write content in many different template processors. These template languages let you add basic logic and variables to HTML. Out of the box, Eleventy supports Markdown, Liquid, Nunjucks, Handlebars, Mustache, EJS, HAML, and Pug. You can also use basic HTML or even JavaScript. The language you pick is up to you, and you can even mix and match different ones in an Eleventy project. The Eleventy docs (https://www.11ty.dev/docs/) won't help you learn these languages, so you'll need to visit the respective web page of the processor you want to use if you aren't familiar with it. For this chapter we'll be using Markdown and Liquid. The choice of Liquid is purely a personal decision; if you don't care for it, remember you've got options!

2.1.1 Creating your first Eleventy site

Let's start by creating a simple site to test Eleventy. Create an empty folder (or clone the book repository) and add a new file, index.md. The "md" extension stands for Markdown, a very common template language used in many different platforms. You can find a good primer on Markdown at https://www.markdownguide.org/.

Listing 2.1 Initial Markdown file, index.md

```
## Hello World

This is my first page in lovely markdown.

Here's a list for fun:

* Cats
* Dogs
* More Cats
```

Now create another file in the same directory, cats.md.

Listing 2.2 The second Markdown file, cats.md

```
## Cats

Cats are the best, aren't they?
```

In case it isn't obvious, you should feel free to change this text to whatever you want. After saving both files, execute a build by typing `eleventy` in your terminal. Ensure you are in the same directory as the two files. Eleventy will report on what it's done as well as tell you how long it took. You can see an example of this in figure 2.3.

```
ray@Mandalore:~/projects/the-jamstack-book/chapter2/demo1$ eleventy
Writing _site/cats/index.html from ./cats.md.
Writing _site/index.html from ./index.md.
Wrote 2 files in 0.07 seconds (v0.11.0)
ray@Mandalore:~/projects/the-jamstack-book/chapter2/demo1$
```

Figure 2.3 Output from running the Eleventy command

There are a few important things to note here. First, Eleventy outputs its result in a directory named _site. You can customize this with a flag passed to the Eleventy command line program. Second, notice how Eleventy converted the file names. The first file, index.md, is written out as `index.html` in the root of the output directory. The second file, cat.md, is written out into a cats subdirectory. The net result is that the home page is available at the root of your site and the cats page is available at `/cats`.

Eleventy gives you full control over how pages are written out, so you can tweak these options to match your needs. However, it's important to know the defaults so that you can link from one page to another. For example, the index page could link to the cats page like so:

```
Read about our [cats](/cats)
```

You can fire up a web server (my preference is httpster, installed via npm) to look at the output, but let's use the other main feature of the Eleventy CLI instead and start the development server. In your terminal, run `eleventy --serve`. You should see output similar to that in figure 2.4.

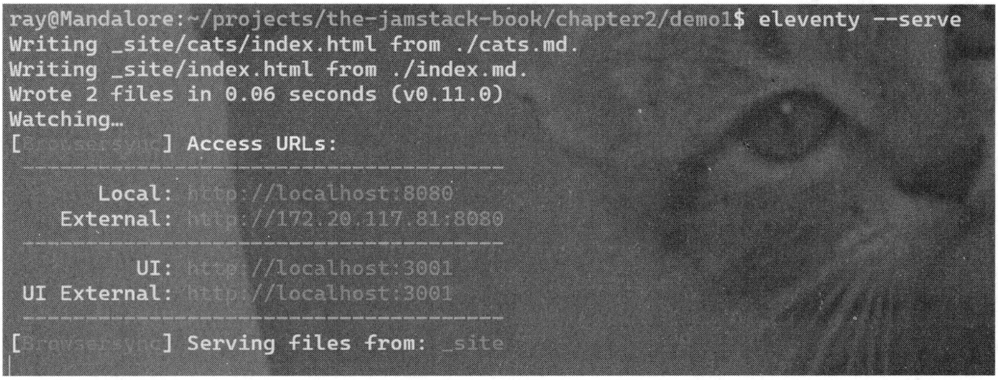

```
ray@Mandalore:~/projects/the-jamstack-book/chapter2/demo1$ eleventy --serve
Writing _site/cats/index.html from ./cats.md.
Writing _site/index.html from ./index.md.
Wrote 2 files in 0.06 seconds (v0.11.0)
Watching…
[Browsersync] Access URLs:
 --------------------------------------
       Local: http://localhost:8080
    External: http://172.20.117.81:8080
 --------------------------------------
          UI: http://localhost:3001
 UI External: http://localhost:3001
 --------------------------------------
[Browsersync] Serving files from: _site
```

Figure 2.4 Running the Eleventy server starts up a web server for your site's files and lets you verify output in your browser.

If you copy the local URL output by the CLI you can open it in your browser and see your site. You will need to press CTRL+C to stop the server. Also note that Eleventy is writing to `_site`, which means when you're done developing, you don't need to run `eleventy` again to generate the final result. If you created your Markdown file exactly as specified, you'll see what figure 2.5 depicts in your browser.

If you edit the file and reload your browser, you'll see the change reflected. Eleventy supports hot reloading, which means you won't have to manually reload your browser, but that feature requires a "full" HTML page that includes body tags. When we get to layouts you'll see how to add that.

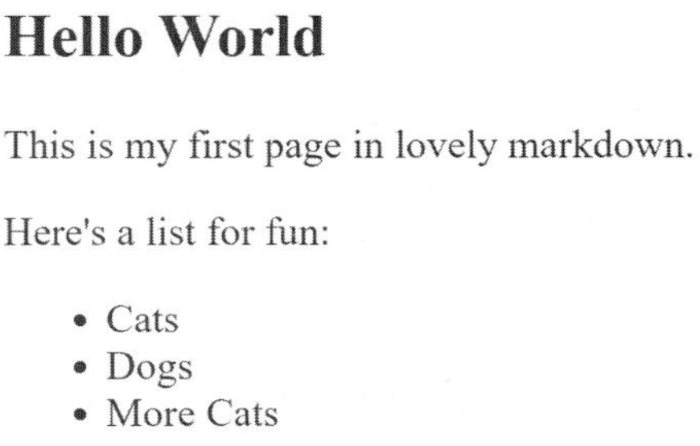

Hello World

This is my first page in lovely markdown.

Here's a list for fun:

- Cats
- Dogs
- More Cats

Figure 2.5 The HTML output generated from the Markdown file processed by Eleventy

2.1.2 *Working with template languages*

So far all you've seen is Markdown conversion, which by itself isn't that exciting. What if we add in a template language? Template languages give us more flexibility in building out HTML (and other) files. They let you use variables, conditionals, and looping. Let's build out an example of this so you can see it in action. You'll see how to add logic and programming statements in an HTML file so that Eleventy can process them and make them static. Create a new directory (or use the chapter2/demo2 folder from the GitHub repository) and add the following as `index.liquid`.

Listing 2.3 An example of the Liquid template language

```
<h2>Liquid Demo</h2>

{% assign name = "ray" %}

<p>
Hello, {{ name }}!
</p>

{% assign cool = true %}

{% if cool %}
<p>
Yes, you are cool.
</p>
{% endif %}

{% assign cats = "Fluffy,Muffy,Duffy" | split: ',' %}
<ul>
{% for cat in cats %}
    <li>{{ cat }}</li>
{% endfor %}
</ul>
```

Liquid (https://shopify.github.io/liquid/) is an open source template language used in many projects, including another static site generator that we'll cover later in the book, Jekyll. In listing 2.3 you can see a bit of the syntax Liquid uses. The first line (ignoring the HTML) assigns a value (`"ray"`) to a variable (`name`). Immediately after the assignment, the variable is output. Liquid uses `{% ... %}` tokens to wrap commands and uses `{{ ... }}` to wrap variables. Basically, if you are applying any kind of logic, use `{%`; otherwise use `{{`.

The next set of code assigns a Boolean value (`true`) to a variable (`cool`), and then the next block checks to see if it's true. Finally, an array is created from a list of names that's then iterated over and displayed in a list. If that syntax looks a bit wonky, keep in mind that these template languages are not supposed to replace a "proper" full development language. Rather, they provide some basic functionality to let you create dynamic HTML.

If you were still running the Eleventy command line, use CTRL+C to stop it, then run `eleventy --serve` from the new directory, and you should see output similar to figure 2.6. Again, you should feel free to modify the values a bit to see them change.

Liquid Demo

Hello, ray!

Yes, you are cool.

- Fluffy
- Muffy
- Duffy

Figure 2.6 Output from the Liquid template

Eleventy used the Liquid processor to take your input and output it as pure HTML. We only looked at a small part of what you can do in Liquid, but remember that if you don't care for this style of writing, Eleventy gives you multiple other options. Before I used Liquid, I was a fan of Handlebars, but I found it limiting at times. Handlebars doesn't want you using a lot of logic in your templates and pushes you to do that elsewhere. I get the logic behind that thinking, but at the same time I prefer the flexibility of Liquid. Liquid can't do everything, so when I need something *really* customized, I'll use EJS (https://ejs.co/). I really don't like how EJS looks. It's a purely personal opinion, but I find EJS code ugly. However, it's easily the most flexible option, so I appreciate that I can use it when I need it.

2.1.3 Adding layouts and includes

Thus far you've seen how Eleventy can transform a Markdown or other language file into HTML. Now it's time to improve this a bit by working with layouts and includes. *Layouts* are "wrapper" files that can be used to add a site layout to your files. Each layout will have a marker where your page content is included. This means you can have one file for your layout and then have the rest of your site use it. Hopefully the power of that is apparent. You can quickly change the entire look and feel of your site just by modifying that one file. You can also correct things like typos in your site more quickly when they're in the header or footer. *Includes* are simply other files that you include in your templates. If you've got a few forms on your site that need to include the same legal text (that readers ignore), includes are a good way to simplify that process.

To begin working with layout files, we first need to tell our templates to use them. Eleventy uses a feature that many static site generators support: front matter. Front matter is metadata about your file included at the top. In almost all cases, front matter uses a particular marker to start and end the block and then a simple "key: value" format to specify values. Here's an example of how front matter may look for a particular file.

Listing 2.4 An example of front matter

```
---
something: a value!
anotherSomething: another value
---

This is the rest of the file.
```

In this incredibly simple example, there are two values specified in front matter, `something` and `anotherSomething`. Eleventy, or any other static site generator, will typically look at these values and do different things based on what you've set. For values they don't understand, the values will be ignored but available to your code. (You'll see this later in the section on data.) When the file is generated to HTML, the front matter is completely removed. Let's look at an example of this.

Create a new directory (or work with your copy of the repository) named demo3. In this directory, copy the index.md and cats.md files from the first demo. Open the first file and modify it to add front matter.

Listing 2.5 Updated index.md file with front matter

```
---
layout: main
---

## Hello World

This is my first page in lovely markdown.

Here's a list for fun:

* Cats
* Dogs
* More Cats
```

The change here is the addition of the front matter block on top, with one value specified: `layout`. The cats.md file can also be similarly modified.

Listing 2.6 Updated cats.md file

```
---
layout: main
---

## Cats

Cats are the best, aren't they?
```

When Eleventy parses these files, it will read in the front matter and notice the layout setting. This will tell Eleventy to try to load a layout file named main dot something. Why "dot something"? As stated previously, Eleventy supports many different template formats with different extensions. By specifying just "main" in the front matter, Eleventy will check for any of the supported template types. Where does it check? By default, layout files are searched in a directory named _includes. This is in the root of your project and the same directory you run the Eleventy command line. As you are working in the demo3 folder, the _includes folder would go under there. You can, if you choose, change this directory name. Let's use Liquid to build a layout file and store it in _includes/main.liquid. Listing 2.7 demonstrates the layout file we built.

Listing 2.7 The site layout file

```
<html>

<head>
<title>My Site</title>
<style>
body {
    background-color: #ffaa00;
    margin: 50px;
}

footer {
    background-color: #c0c0c0;
    padding: 10px;
}
</style>
</head>

<body>

    {{ content }}    <-- This is where page content will be included.

    <footer>
    <a href="/">Home</a> | <a href="/cats">Cats</a>
    </footer>
</body>
</html>
```

For the most part, this looks like a regular HTML template. You can see some styling (ugly, to be sure) and a footer, but make special note of `{{ content }}`. When Eleventy parses a file that uses front matter and that specifies this layout, it will take the resulting HTML generated by the page and place it in a variable named `content`. My Liquid template can then simply output that variable wherever it makes sense. In this case I've done it right after the `body` tag and before the `footer`. To be clear, this is arbitrary. If you fire up an Eleventy server in this directory (remember to quit any previous test by pressing CTRL+C), you'll see in figure 2.7 why I'm not allowed to design websites.

If you want, try changing the style declarations and pick different (better) color choices. Now that you've got a complete HTML page being used you'll finally see Eleventy's hot reload feature in action.

Before leaving the topic of layouts, note that you can also have layouts that have layouts. For example, instead of specifying a main layout for a page, you could specify another one that then specifies the main layout itself. The first one will run, include your page contents, and then return its HTML to the main layout. You'll see an example of this later.

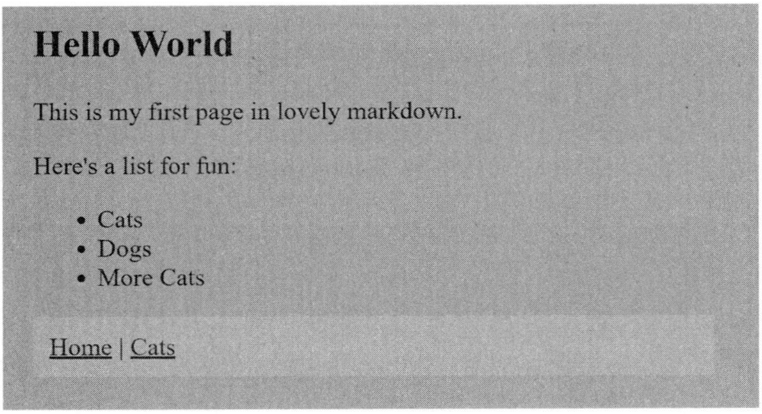

Hello World
This is my first page in lovely markdown.
Here's a list for fun:
• Cats
• Dogs
• More Cats
Home | Cats

Figure 2.7 Our site with the layout applied

Now let's demonstrate includes. This is typically rather simple. You create a file to be included and store it in the _includes folder where your layouts reside as well. How you include the file depends on your template engine. For Liquid, it looks like this:

```
{% include footer %}
```

where `footer.liquid` should be found in the _includes folder. For Handlebars, it looks a bit different:

```
{{> footer}}
```

In Handlebars, these are referred to as *partials*, but they act the same way. Eleventy has documentation (https://www.11ty.dev/docs/languages/) for all its supported template languages, so check to see how this feature is supported in your template language of choice.

2.1.4 Using collections in your Eleventy site

Thus far you've seen how to "translate" input files, in Markdown and other languages, into HTML. Now it's time to demonstrate another powerful feature—collections. A collection is exactly how it sounds: a set of files that are grouped together by some logical pattern. In Eleventy, there's a few different ways of doing this, but the simplest way is via front matter and using the `tags` value. Consider the following front matter example:

```
---
layout: main
tags: pressReleases
title: Press Release One
---
```

Here, I've used three front matter values. The first, `layout`, we've already discussed. The second, `tags`, specifies a tag value for this file. Any file that uses the same value will be in the same collection. You can choose multiple tags for a file if you want, and it will be available in multiple collections. The last value, `title`, is one I haven't shown before. As you can guess, this specifies a title for your file and will be available in your templates to output if you choose.

By default, this won't do anything on its own. But your templates can iterate over collections and create lists. Let's build an example that works with a collection and creates a list of them dynamically.

Create a new directory named demo4 (or use the source from the repository). This should not be in the same folder as demo3. Copy the same contents from demo3 and rename index.md to index.liquid. We're going to add press releases to our site, so add a subdirectory named press-releases. In this directory, create a few files. The names aren't terribly important, nor is the content, but if you want to follow along with the repository, name the first one cats-are-cool.md.

Listing 2.8 The first press release

```
---
layout: pr
tags: pressReleases
title: Cats are Cool
---

Just some text here for filler.
```

As we discussed, we're using our front matter to specify a tag for the file to add it to a collection. We've also specified a layout and title. Repeat this a few times (the repository has three files), but the names, titles, and content don't really matter. The repository has a file named dogs-are-not-cool.md and have-we-said-how-cool-cats-are.md.

The next step is to add the layout. We mentioned earlier that layouts can include other layouts. For our press releases, we want them to have some additional layout.

Listing 2.9 The "pr" layout in _includes/pr.liquid

```
---
layout: main
---

<h2>Press Release: {{ title }}</h2>    <-- Outputting the title

{{ content }}
```

Notice that it specifies `main` for its layout. This means the press release will first run the pr.liquid template and then send that output to main.liquid. Also notice the title is output in the layout. This value comes from the title in the front matter of the press release. We can also make use of this in our core layout file.

Listing 2.10 Updated layout file supporting a title (_includes/main.liquid)

```
---
title: Default title
---

<html>

<head>
<title>{{ title }}</title>
<style>
body {
    background-color: #ffaa00;
    margin: 50px;
}

footer {
    background-color: #c0c0c0;
    padding: 10px;
}
</style>
</head>

<body>

    {{ content }}

    <footer>
    <a href="/">Home</a> | <a href="/cats">Cats</a>
    </footer>
</body>
</html>
```

There are only two changes in this version of the file. First, a title value is specified in the front matter. This will only be used if a template doesn't specify its own title. Next, the title is output between the title tags.

The end result is that our press release files will automatically get a bit of additional layout and then use the main layout the rest of the site has. Now that we've got press releases and we know they're in a collection, the next step is to expose this to the user by creating a new home page that uses the Liquid template engine to output these values.

Listing 2.11 The index.liquid with press release support (/index.liquid)

```
layout: main
title: Home Page

Welcome to our home page. Here's our press releases:
```

```
<ul>
{% for pr in collections.pressReleases %}
<li><a href="{{ pr.url }}">{{ pr.data.title }}</a></li>
{% endfor %}
</ul>
```

Note the use of collections.pressReleases.

When Eleventy encounters templates using tags in their front matter, it puts them in an object called `collections`. You can then access that data by using `collections.nameoftag`. Since we used `pressReleases` as the tag value, this collection will consist of three items. (Again, feel free to make more press releases and see the changes.) We loop over each item and assign the object to a `pr` variable. That variable has multiple different properties in it. One is `url`, which is created by Eleventy and represents the location of the page. The `data` property will contain any data defined in the front matter of the page. Since our press releases have titles, we can output them here. There are more variables you can use, and you can check the collections documentation (https://www.11ty.dev/docs/collections/) for details. The end result is shown in figure 2.8.

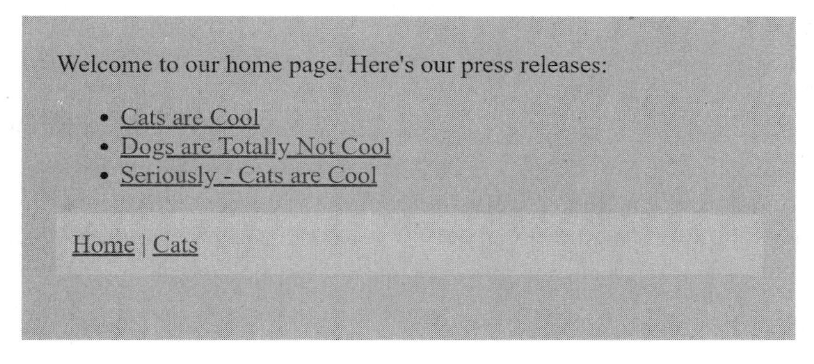

Welcome to our home page. Here's our press releases:
Cats are Cool
Dogs are Totally Not Cool
Seriously - Cats are Cool
Home | Cats

Figure 2.8 Our home page with press releases driven by the Eleventy collection

If you click on one of the press releases, you can see the multilayered layout in effect (figure 2.9).

Figure 2.9 A sample press release

Now that you've seen some examples of using front matter and collections, it's time to look at data in Eleventy at a high level.

2.1.5 Working with data

You've had a small introduction to how data works in Eleventy, but now we're going to dig a bit deeper. There are multiple ways of providing data to Eleventy pages. Front matter is one way and is specific to a page. Collections are another way and provide data that can be iterated and used anywhere. Another way to provide data to Eleventy sites is by creating files with a special folder: _data. You can create two types of files here, JSON and JavaScript. The name of the file will define how it's available to your site. For example, imagine a file named site.json within your _data folder. It has the following contents:

```
{
"siteName": "The Cat Site",
"siteOwner": "Raymond Camden",
"siteEmail": "raymondcamden@gmail.com"
}
```

This JSON file defines three variables. It could also be an array, or an array of objects—basically any valid JSON. Because the file is named site.json, your templates can access values like so (again, assuming Liquid syntax): `{{ site.siteName }}`.

JavaScript data files work a bit differently because, of course, JavaScript is a programming language, and JSON is just static information. JavaScript data files let you perform any operation you may need, and when done the result of the file is the final available data. Imagine a sample JavaScript file named build.js:

```
module.exports = function() {
    let buildTime = new Date();
    let randomNumber = Math.random();
    return {
        buildTime,
        randomNumber
    }
}
```

This JavaScript file creates two variables, one representing the current time and one a random number. An object containing both values is then returned. Because the file is named build.js, our templates can use the values like `{{ build.buildTime }}` or `{{ build.randomNumber }}`. Note that this logic is evaluated when the site is built (the file name doesn't have anything to do with that), so when the static site is deployed, the values won't change. Let's look at an example of both kinds of data files in action.

Begin by once again copying the previous directory (demo4) to a new one (demo5), or simply using the GitHub repository code. Create a new folder inside demo5 named _data, which will store the data files. The folder name used for data files can be configured in Eleventy if you wish. The first file we'll add is site.json.

Listing 2.12 The site.json definitions (_data/site.json)

```
{
    "siteName":"The Cat Fan Site",
    "contactAddress":"raymondcamden@gmail.com"
}
```

The values are arbitrary and hopefully self-explanatory. We've defined a name for the site as well as a contact address. Now define a new file, breeds.js. For this file we're going to make use of the Cat API site (https://thecatapi.com/). This site provides free APIs related to cats, but you have to sign up (https://thecatapi.com/signup) for a key first as shown in figure 2.10.

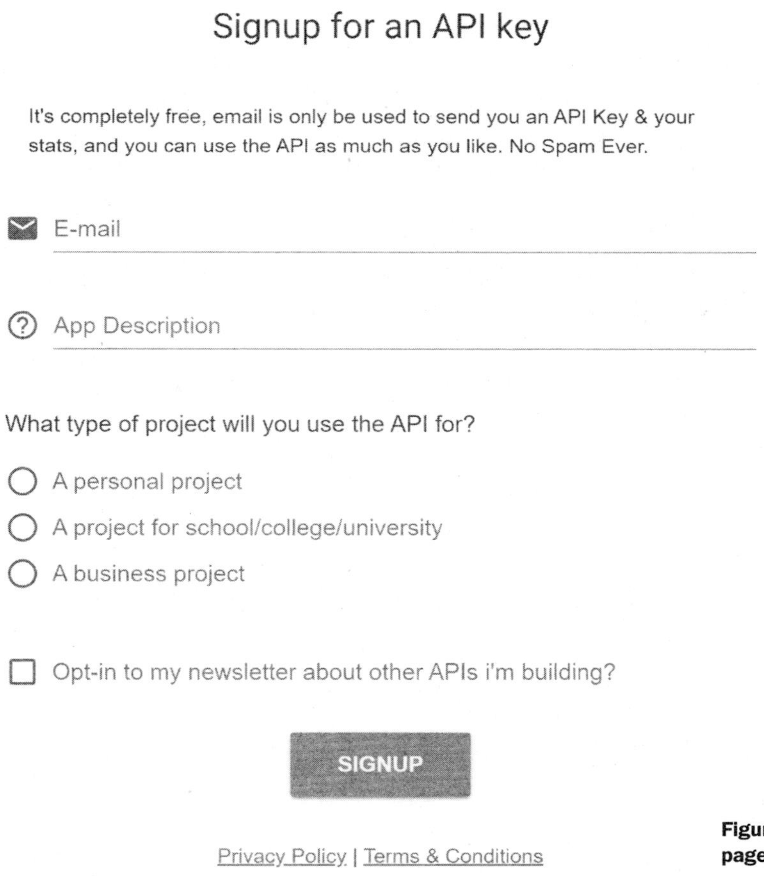

Signup for an API key

It's completely free, email is only be used to send you an API Key & your stats, and you can use the API as much as you like. No Spam Ever.

E-mail

App Description

What type of project will you use the API for?

A personal project

A project for school/college/university

A business project

Opt-in to my newsletter about other APIs i'm building?

SIGNUP

Privacy Policy | Terms & Conditions

Figure 2.10 The sign-up page at the Cat API

One of the APIs they provide is to return a list of cat breeds. The API is rather simple and can be called with this URL: https://api.thecatapi.com/v1/breeds?limit=5&api_key=yourkey. The limit value in the URL restricts the amount of data returned, and the key value should be replaced with the one you get.

Listing 2.13 Retrieving a list of cat breeds (_data/breeds.js)

```
const fetch = require('node-fetch');
require('dotenv').config();            <-- Loading the 'dot.env' package

const KEY = process.env.CAT_API_KEY;     <-- Reading in the .env key value

module.exports = async function() {

    let breedUrl = 'https://api.thecatapi.com/v1/breeds?limit=5' +
    '&api_key=' + KEY;

    let resp = await fetch(breedUrl);

    let data = await resp.json();

    return data;
}
```

In general, this is a simple Node script that makes use of node-fetch to make easier HTTP calls. The body of the function makes the call and returns the results as-is. (It could be modified to transform the code, too.) The important bit here is the use of the `'dot.env'` package on line 2. This will look for a file named .env in the root of your project. This file is a simple set of name value pairs. You won't find this in the GitHub repository as it's a standard way of including keys in your code. The Node script will read this and set every value equal to a value in the `process.env` scope. In the .env file at the root of the demo5 folder, add the following:

```
CAT_API_KEY=yourkey
```

Replace `yourkey` with the key you got from the Cat API.

There's one last step to do. Our function made use of node-fetch and `'dotenv'`. In order to make those available, you'll need to install them via npm. In your terminal, in the same directory as the demo5 folder, run

```
npm install node-fetch@2
npm install dotenv
```

This will install the required dependencies.

Now that we've set up the data files, let's use them. To keep it simple, let's use them both in index.liquid.

Listing 2.14 Making use of Eleventy data files

```
---
layout: main
title: Home Page
---
```

```
Welcome to our home page. Here's our press releases:

<ul>
{% for pr in collections.pressReleases %}
<li><a href="{{ pr.url }}">{{ pr.data.title }}</a></li>
{% endfor %}
</ul>

Here's a list of cat breeds:

<ul>
{% for breed in breeds %}
<li>{{ breed.name }} - {{ breed.description }}</li>
{% endfor %}
</ul>

You can contact me at {{ site.contactAddress }}
```

The modifications begin right after the press release display, though note that the logic is pretty much the same: loop over the value and output data from each item. The Cat API returns a lot of data about cat breeds, but to keep it simple, we've output the name and description. Finally, the site contact address is displayed at the end. Figure 2.11 shows the result.

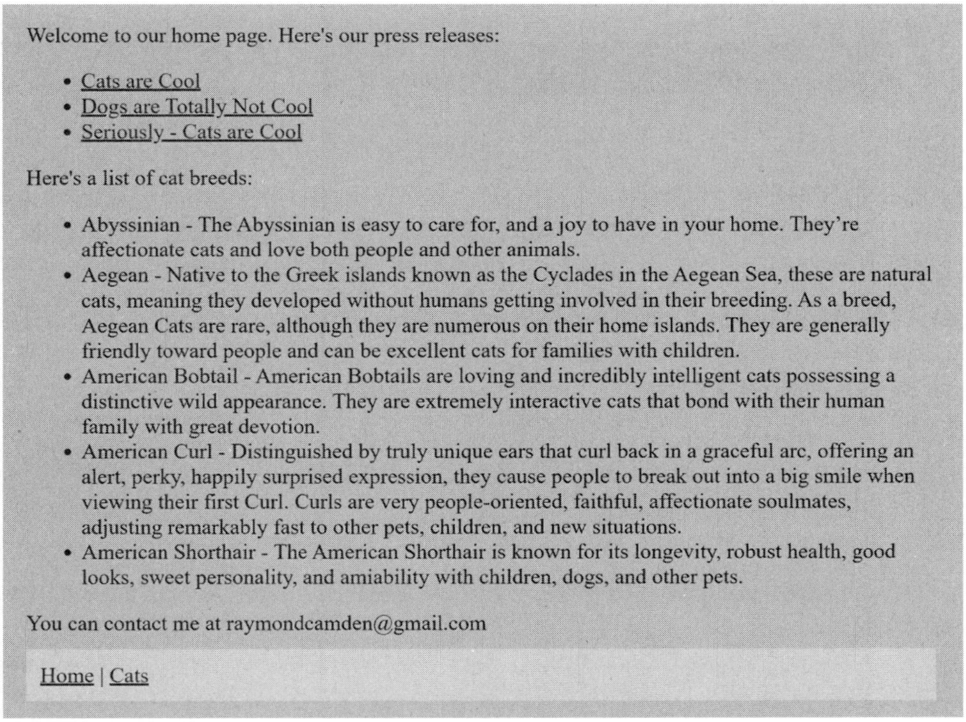

Welcome to our home page. Here's our press releases:
- Cats are Cool
- Dogs are Totally Not Cool
- Seriously - Cats are Cool
Here's a list of cat breeds:
- Abyssinian - The Abyssinian is easy to care for, and a joy to have in your home. They're affectionate cats and love both people and other animals.
- Aegean - Native to the Greek islands known as the Cyclades in the Aegean Sea, these are natural cats, meaning they developed without humans getting involved in their breeding. As a breed, Aegean Cats are rare, although they are numerous on their home islands. They are generally friendly toward people and can be excellent cats for families with children.
- American Bobtail - American Bobtails are loving and incredibly intelligent cats possessing a distinctive wild appearance. They are extremely interactive cats that bond with their human family with great devotion.
- American Curl - Distinguished by truly unique ears that curl back in a graceful arc, offering an alert, perky, happily surprised expression, they cause people to break out into a big smile when viewing their first Curl. Curls are very people-oriented, faithful, affectionate soulmates, adjusting remarkably fast to other pets, children, and new situations.
- American Shorthair - The American Shorthair is known for its longevity, robust health, good looks, sweet personality, and amiability with children, dogs, and other pets.
You can contact me at raymondcamden@gmail.com
Home | Cats

Figure 2.11 The updated home page driven by dynamic data files

Be sure to check the "Using Data" section (https://www.11ty.dev/docs/data/) of the Eleventy docs for more examples. Eleventy also supports specifying data at a directory level, as well as other options.

2.2 Let's build Camden Grounds

Now that we've seen a bit (certainly not all!) of what Eleventy can do, it's time to look into building our simple restaurant site. The site we will be building in this chapter is for a fictional coffee shop named Camden Grounds. This will follow the typical pattern of most restaurant sites by offering a few basic features:

- A home page that's mostly just pretty pictures to help entice the visitor.
- A menu of product offerings, again with pretty pictures.
- A Locations page showing where different Camden Grounds stores may be found.
- An About Us page that talks about the history of the store. No one will ever actually read this, but the owner insists on it.

To begin, you need to determine the design of the site. If you're like me and design challenged, the simplest solution is to find a template you can use. Luckily, there's a large number of websites that cater to this need. We're going to use a simple Bootstrap store template named "Shop Homepage" (http://mng.bz/Dxy0). You can see a preview of this in figure 2.12. Note that this is how it looks in its template form, not the final form we'll be creating.

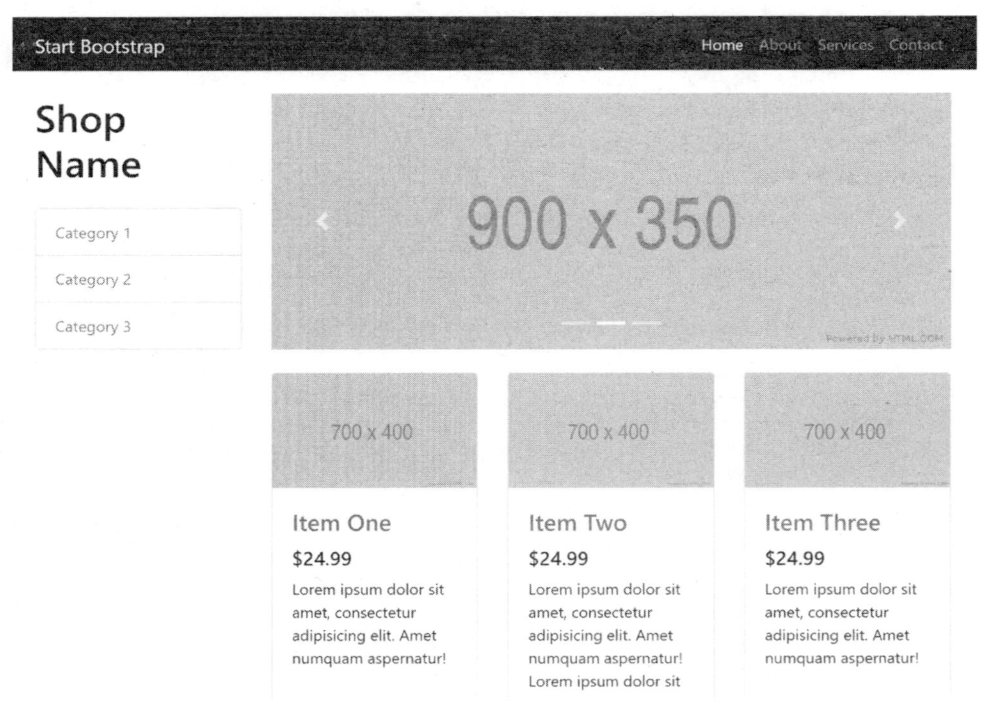

Figure 2.12 The shop template with products and a rotating banner on top

NOTE Since this chapter was created, the template updated to a newer version with a slightly different design. The version used in this chapter is still available at http://mng.bz/2j6w and can be used. Also remember that the template can be found in the completed code for the book at http://mng.bz/1j5R.

This template can be freely downloaded and modified. The full source code of the final demo can be found in the GitHub repository in the chapter2/camdengrounds folder. As there's a lot of boilerplate HTML (Bootstrap is nice, but a bit verbose), we'll be focusing on the modifications of the template rather than sharing every line of code.

NOTE While a template may look pretty, before you commit, look at the code. It may be difficult to modify for your needs.

As stated earlier, you can think of a typical template as a "wrapper" around your pages. You figure out where your content can be stuffed, insert the token that Eleventy will use to put page content in, and then ensure your pages specify that layout.

You can see the template for Camden Grounds in the _includes/main.liquid file. As there's a lot of HTML in this file, the next few listings will highlight the important parts. First, the very top of the template specifies a title in front matter.

Listing 2.15 Front matter for the layout

```
---
title: Camden Grounds
---
```

This will be used as a default when individual pages do not specify one. You can see this a few lines later.

Listing 2.16 The dynamic title

```
<title>{{ title }}</title>
```

Now let's look at the navigation menu. One of the features this template supports is the ability to highlight a tab when you're on a particular page. How do we use that on a static website? Eleventy provides access to a `page` variable that provides information about the current page being rendered. One of the values is the relative URL of the page. Given that we can use a template language and examine that value, we can create a dynamic menu.

Listing 2.17 The template's dynamic menu

```
<ul class="navbar-nav ml-auto">
    <li class="nav-item {% if page.url == '/' %}active{% endif %}">
    <a class="nav-link" href="/">Home
```

```
        </a>
        </li>
        <li class="nav-item {% if page.url == '/about/' %}active{% endif %}">
        <a class="nav-link" href="/about">About</a>
        </li>
        <li class="nav-item {% if page.url == '/services/' %}active{% endif %}">
        <a class="nav-link" href="/services">Services</a>
        </li>
        <li class="nav-item {% if page.url == '/contacts/' %}active{% endif %}">
        <a class="nav-link" href="/contact">Contact</a>
        </li>
</ul>
```

Our navigation consists of four main pages. For each page, Liquid code is used to look at the URL value, and when it matches the relative path of the page in question, the `active` class is added. The last part of the template we'll share is where the contents are included.

Listing 2.18 Including page content in the template

```
<div class="col-lg-9">

{{ content }}

</div>
```

To be honest, it took a bit of digging to find the optimal place to add the `content` variable. As stated, Bootstrap is a great design framework but can be both verbose and a bit complex at times. You are highly encouraged to keep a tab open at the official Bootstrap site (https://getbootstrap.com/). It has extensive documentation and examples to help you better use the framework.

The home page is going to be a bit complex, so let's look at the three simple pages, one by one.

Listing 2.19 The About page (about.html)

```
---
layout: main
title: About Camden Grounds
---

<div class="row my-4">
    <div class="col">

    <h2>The About Page</h2>
    <p>
    Let's talk about the site.
    </p>

    </div>
</div>
```

If you look at the source of the layout file, you'll see it's nearly 100 lines long. This template is far shorter, as Eleventy will handle wrapping it with the content. You can look at `contact.html` and `services.html` if you wish; they follow the exact same format. Figure 2.13 shows the how the About page displays (notice the highlighted menu tab).

Figure 2.13 The Camden Grounds About page

Before this will work, though, it's time to point out an interesting issue with Eleventy. When it encounters a file it doesn't know how to process, it ignores it. This ends up being a big deal as our Bootstrap template has multiple folders of CSS, JavaScript, and images that make up the look and feel of the design.

What we need to do is tell Eleventy to copy these files as-is. Eleventy supports dynamic configuration by specifying a .eleventy.js file in your project root. While you can do a lot in this file, the main feature we want to enable is called Passthrough File Copy (https://www.11ty.dev/docs/copy/). In the root of the Camden Grounds site is the following configuration.

Listing 2.20 The Eleventy configuration file (.eleventy.js)

```
module.exports = function(eleventyConfig) {

  eleventyConfig.addPassthroughCopy("css");
  eleventyConfig.addPassthroughCopy("vendor");
  eleventyConfig.addPassthroughCopy("img");

};
```

We've specified that three paths, `css`, `vendor`, and `img`, should be copied as-is. Eleventy won't try to process anything here, but will simply copy the files, recursively, into the generated _site folder.

All right, we've done the easiest bits first (always a good idea), but now it's time to tackle the home page. If you look back at figure 2.12, you'll see that the home page has a rotating banner of images on top and then a list of images. If you look at the source of the index page in the template's original code, you'll see that the animated images (called a *carousel*) are displayed.

Listing 2.21 Image carousel code

```
<div class="carousel-inner" role="listbox">
<div class="carousel-item active">
    <img class="d-block img-fluid" src="http://placehold.it/900x350"
    alt="First slide">
</div>
<div class="carousel-item">
    <img class="d-block img-fluid" src="http://placehold.it/900x350"
    alt="Second slide">
</div>
<div class="carousel-item">
    <img class="d-block img-fluid" src="http://placehold.it/900x350"
    alt="Third slide">
</div>
</div>
```

You can see each image is using the Placeholder service to display a temporary image. We're going to replace this with three images from Unsplash (https://unsplash.com/). Unsplash provides free images for use on websites with the simple request that you attribute them. I searched for "coffee" and found three images and placed them in the img folder.

Listing 2.22 The image carousel with our new images specified

```
   <div class="carousel-inner" role="listbox">
    <div class="carousel-item active">
      <img class="d-block img-fluid" src="img/coffee1.jpg"
      alt="Coffee!">
    </div>
    <div class="carousel-item">
      <img class="d-block img-fluid" src="img/coffee2.jpg"
      alt="Coffee beans">
    </div>
    <div class="carousel-item">
      <img class="d-block img-fluid" src="img/coffee3.jpg"
      alt="Coffee cup">
    </div>
  </div>
```

Now for the products. The original template shows six products using what's commonly referred to as a *card* format. We want to make that dynamic, so let's add a JSON file with products in _data/products.json.

Listing 2.23 The products for Camden Grounds

```
[
    {
        "name" : "Coffee",
        "price" : 2.99,
        "description" : "Lorem ipsum dolor sit amet, consectetur adipisicing
         ➥ elit. Amet numquam aspernatur!",
```

```
        "thumbnail" : "http://placehold.it/700x400",
        "image" : "http://placehold.it/900x350"
    },
    {
        "name" : "Espresso",
        "price" : 3.99,
        "description" : "Lorem ipsum dolor sit amet, consectetur adipisicing
        ➥ elit. Amet numquam aspernatur!",
        "thumbnail" : "http://placehold.it/700x400",
        "image" : "http://placehold.it/900x350"
    },
    {
        "name" : "Americano",
        "price" : 5.99,
        "description" : "Lorem ipsum dolor sit amet, consectetur adipisicing
        ➥ elit. Amet numquam aspernatur!",
        "thumbnail" : "http://placehold.it/700x400",
        "image" : "http://placehold.it/900x350"
    },
    {
        "name" : "Double Espresso",
        "price" : 8.99,
        "description" : "Lorem ipsum dolor sit amet, consectetur adipisicing
        ➥ elit. Amet numquam aspernatur!",
        "thumbnail" : "http://placehold.it/700x400",
        "image" : "http://placehold.it/900x350"
    },
    {
        "name" : "Tea",
        "price" : 1.99,
        "description" : "For those who prefer tea.",
        "thumbnail" : "http://placehold.it/700x400",
        "image" : "http://placehold.it/900x350"
    }
]
```

Note that the text was slightly reduced for space. Instead of finding a custom picture for each product, we use an image placeholder service from http://placehold.it that produces a basic gray image that is useful for temporary images. Now that this has been set, Eleventy provides access to them in a `products` variable. We can modify the index template (camdengrounds/index.html) to loop over and create a card for each product.

Listing 2.24 Displaying products

```
{% for product in products %}
<div class="col-lg-4 col-md-6 mb-4">
<div class="card h-100">
    <a href="products/{{ product.name | slug }}"><img class="card-img-top"
    src="http://placehold.it/700x400" alt=""></a>
    <div class="card-body">
    <h4 class="card-title">
```

```
        <a href="products/{{ product.name | slug }}">{{ product.name }}</a>
    </h4>
    <h5>${{product.price}}</h5>
    <p class="card-text">{{product.description}}</p>
    </div>
</div>
</div>
{% endfor %}
```

Basically all we did was remove the boilerplate text and replace it with variables defined by the loop itself. If we had unique pictures for our products (hopefully), they could also be specified in the JSON file and used here.

The last aspect of the site is the pages for each individual product. You can see the links used in listing 2.25: `products/{{ product.name | slug }}`. The first thing to note is the `slug` aspect. This is called a *filter* and takes the input provided (`product.name`) and then passes it through a formatting function. The `slug` function creates a URL-safe version of a string. As an example of this, the value "Double Espresso" turns into "double-espresso." But where do the pages themselves come from?

Eleventy supports a powerful Pagination system (https://www.11ty.dev/docs/pagination/) that can be used in multiple ways. It can take a list of data and create pages (e.g., hundreds of press releases split into pages of 10 each) or take a list and generate one page each. Eleventy also provides support for linking to pages, either the next or previous one, and even makes it easy to tell if you are at the beginning of a list of pages or the end. As with most things in Eleventy, it's pretty powerful and flexible, but we're going to use a simple example of it in our shop to help create one page for each of our products.

To use this feature, a page must define pagination settings in the front matter, as shown in the following listing.

Listing 2.25 The Products pagination page

```
---
layout: main
pagination:
    data: products
    size: 1
    alias: product
permalink:"products/{{ product.name | slug }}/index.html"
---

<div class="row my-4">
  <div class="col">

      <div class="card mt-4">
            <img class="card-img-top img-fluid"
            src="http://placehold.it/900x400" alt="">
            <div class="card-body">
              <h3 class="card-title">{{ product.name }}</h3>
              <h4>${{product.price}}</h4>
```

```
            <p class="card-text">
            {{ product.description }}
            </p>
          </div>
        </div>

  </div>
```

The front matter makes use of pagination by first specifying the data that will be iterated, then the size (how many per page), and finally that it should use an alias in the page itself for easier use. Using a page size of 1 basically means we want one product per page. Another feature of Eleventy is demonstrated here as well: permalink. Eleventy gives you precise control over how files are generated by letting you specify exactly where they should end up. The rest of the template is basically a modified version of the cards used earlier but could certainly be more unique.

If you fire up the Eleventy server and click on one of the products, you see a product page, as figure 2.14 demonstrates.

Figure 2.14 A Camden Grounds product

2.3 Going further with Eleventy

In this chapter, you've reviewed the Eleventy static site generator and used it to build a simple but flexible shop for a fictional store. However, there are many features in Eleventy we didn't touch upon:

- Filters and shortcodes
- Plug-ins for additional functionality

- Customizing your build with events
- And much more

Be sure to check the Eleventy documentation (https://www.11ty.dev/docs/) for more information. I'd also suggest following the official Twitter account for Eleventy (@eleven_ty) as well as joining the Discord channel (https://www.11ty.dev/news/discord/). Finally, you can peruse the source code and current open issues on their GitHub repository: https://github.com/11ty/eleventy/.

In the next section, you're going to look at another static site generator, Jekyll. Jekyll is tailor-made for blogs, so you'll build a blog as a way to help you learn. Best of all, Jekyll also makes use of Liquid, so if you enjoyed using it in Eleventy, you'll have the opportunity to use it more!

Summary

- Eleventy is a flexible static site generator and a good way to build Jamstack sites.
- Eleventy supports different types of template languages and lets you select one that best matches your style.
- Multiple different template languages can be used in one project if you discover that you need a particular feature from multiple options.
- Simple sites, even ones with only a few pages, are excellent candidates for the Jamstack.

3 Building a blog

> **This chapter covers**
> - Using a static site generator to build a blog
> - How to install and make use of Jekyll
> - Creating a basic blog with Jekyll

Blogs are one of the most popular forms of websites on the internet, with over 75 million people alone using the WordPress platform. A blog is typically fairly simple in structure but highly personal in nature. Much like a diary, a blog lists posts in chronological order, typically presented with the newest first on the home page. Blogs will often have categories and tags as a way to categorize and organize content, letting readers quickly find related content to a particular post.

In this chapter, we will focus on building a blog with the Jamstack. The static site generator we will use builds a blog by default, so a lot of work will be done for us, which leaves us more time to focus on customization of the blog theme (how it looks) and other aspects.

3.1 Blogging with Jekyll

Jekyll (jekyllrb.com) is a static site generator that is specifically tailored toward blogs (figure 3.1). While you can definitely build other types of sites with Jekyll

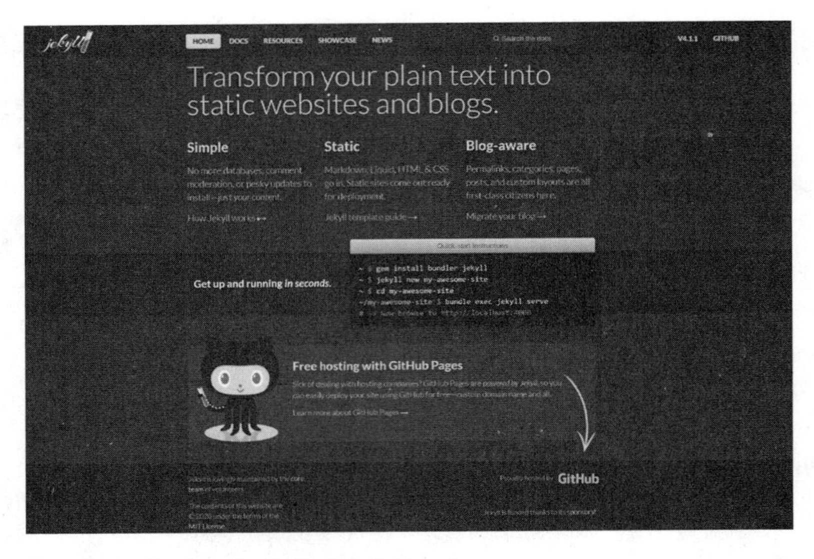

Figure 3.1 The Jekyll website (jekyllrb.com)

(and build blogs with other generators), you'll find Jekyll really shines when building a Jamstack blog. Jekyll is what powers GitHub Pages web pages, which can be incredibly useful for people already using GitHub for their projects.

Installation of Jekyll is somewhat more complex than most static site generators and can be a bit problematic on Windows. (For detailed operating system–specific instructions, see the online documentation at https://jekyllrb.com/docs/installation/.) At this time, Jekyll isn't officially supported on Windows. Personally, I've used Jekyll on Windows for some time, and it works fine, but the lack of support is something to consider before you commit to Jekyll. Modern Windows machines support WSL (Windows subsystem for Linux), which may provide a much smoother experience with Ruby if you have trouble.

Installing Jekyll requires installing Ruby first. Instructions for that can be found at https://www.ruby-lang.org/en/downloads/. You'll then need RubyGems (https://rubygems.org/pages/download). Don't worry if you've never used Ruby before, as it won't be required to create websites with Jekyll. While Jekyll itself runs on Ruby, you will not need to write a line of Ruby unless you start to customize the Jekyll engine itself.

Once you've gotten the prerequisites installed, you can install Jekyll itself with the command:

```
gem install jekyll bundler
```

Note that there are installation guides (https://jekyllrb.com/docs/installation/#guides) for multiple operating systems with more details.

As a final step, ensure Jekyll installed correctly by running `jekyll` in your terminal, as shown in figure 3.2.

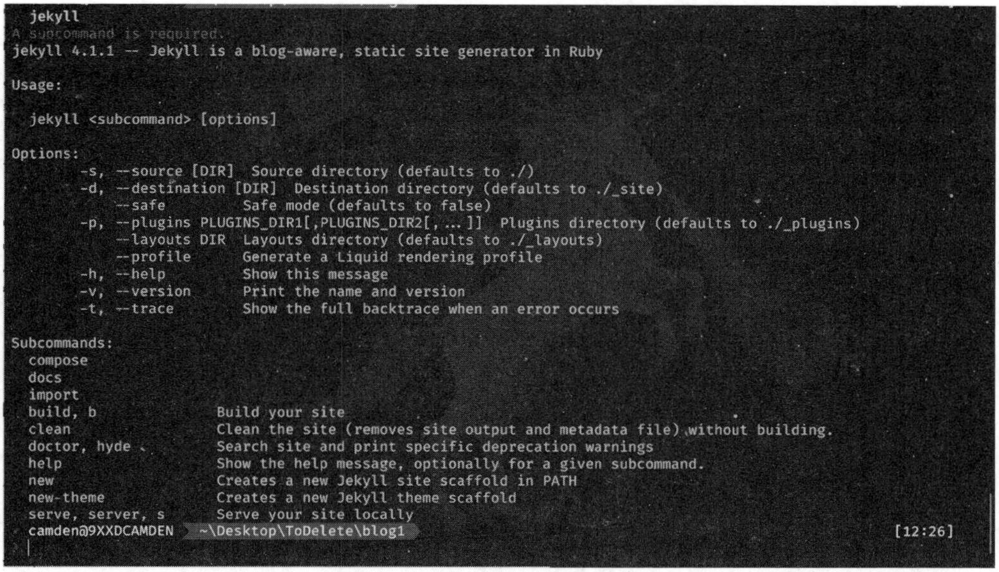

```
 jekyll
A subcommand is required.
jekyll 4.1.1 -- Jekyll is a blog-aware, static site generator in Ruby

Usage:

  jekyll <subcommand> [options]

Options:
        -s, --source [DIR]  Source directory (defaults to ./)
        -d, --destination [DIR]  Destination directory (defaults to ./_site)
            --safe         Safe mode (defaults to false)
        -p, --plugins PLUGINS_DIR1[,PLUGINS_DIR2[,...]]  Plugins directory (defaults to ./_plugins)
            --layouts DIR  Layouts directory (defaults to ./_layouts)
            --profile      Generate a Liquid rendering profile
        -h, --help         Show this message
        -v, --version      Print the name and version
        -t, --trace        Show the full backtrace when an error occurs

Subcommands:
  compose
  docs
  import
  build, b              Build your site
  clean                 Clean the site (removes site output and metadata file) without building.
  doctor, hyde          Search site and print specific deprecation warnings
  help                  Show the help message, optionally for a given subcommand.
  new                   Creates a new Jekyll site scaffold in PATH
  new-theme             Creates a new Jekyll theme scaffold
  serve, server, s      Serve your site locally
  camden@9XXDCAMDEN  ~\Desktop\ToDelete\blog1                                    [12:26]
```

Figure 3.2 Default output from running the `jekyll` command in your terminal

Jekyll supports creating content in HTML and Markdown, but also supports Liquid templates. If you read the previous chapter, you have an introduction to Liquid already, but at a minimum in this chapter, you'll learn how to embed Liquid tags to add logic and dynamic aspects to your templates. Jekyll, like most static site generators, also makes use of front matter as a way to define metadata about a particular page.

3.2 Your first Jekyll site

Jekyll's command line program can help you start development by scaffolding an entire, but small, blog for you out of the box. This includes a layout (with minimal design), a home page (that lists the blog posts), and one sample post. Scaffolding is done by using the command `jekyll new sitename`, where `sitename` should be the name of your site. The value you use for `sitename` will also be used to name the folder for your site. (Note that Jekyll seems to have an issue with spaces in folders, so avoid that when creating your new project.) After you execute the command, Jekyll will create the directory, create some files, and then install various dependencies and other such items needed for it to operate. Figure 3.3 demonstrates how this should look.

Once the command is done running, change directories to the newly created directory. You can test your new Jekyll blog by using the serve command `jekyll serve`. This will start a local application server to process your files and make them available on a local web server. Figure 3.4 demonstrates the output you should see in your terminal.

```
> jekyll new myblog
Running bundle install in C:/Users/camden/Desktop/ToDelete/myblog...
  Bundler: Fetching gem metadata from https://rubygems.org/..........
  Bundler: Fetching gem metadata from https://rubygems.org/.
  Bundler: Resolving dependencies...
  Bundler: Using public_suffix 4.0.5
  Bundler: Using addressable 2.7.0
  Bundler: Using bundler 1.17.2
  Bundler: Using colorator 1.1.0
  Bundler: Using concurrent-ruby 1.1.7
  Bundler: Using eventmachine 1.2.7 (x64-mingw32)
  Bundler: Using http_parser.rb 0.6.0
  Bundler: Using em-websocket 0.5.1
  Bundler: Using ffi 1.13.1 (x64-mingw32)
  Bundler: Using forwardable-extended 2.6.0
  Bundler: Using i18n 1.8.5
  Bundler: Using sassc 2.4.0 (x64-mingw32)
  Bundler: Using jekyll-sass-converter 2.1.0
  Bundler: Using rb-fsevent 0.10.4
  Bundler: Using rb-inotify 0.10.1
  Bundler: Using listen 3.2.1
  Bundler: Using jekyll-watch 2.2.1
  Bundler: Using rexml 3.2.4
  Bundler: Using kramdown 2.3.0
  Bundler: Using kramdown-parser-gfm 1.1.0
  Bundler: Using liquid 4.0.3
  Bundler: Using mercenary 0.4.0
  Bundler: Using pathutil 0.16.2
  Bundler: Using rouge 3.22.0
  Bundler: Using safe_yaml 1.0.5
  Bundler: Using unicode-display_width 1.7.0
  Bundler: Using terminal-table 1.8.0
  Bundler: Using jekyll 4.1.1
  Bundler: Using jekyll-feed 0.15.0
  Bundler: Using jekyll-seo-tag 2.6.1
  Bundler: Using minima 2.5.1
  Bundler: Using thread_safe 0.3.6
  Bundler: Using tzinfo 1.2.7
  Bundler: Using tzinfo-data 1.2020.1
  Bundler: Using wdm 0.1.1
  Bundler: Bundle complete! 6 Gemfile dependencies, 35 gems now installed.
  Bundler: Use `bundle info [gemname]` to see where a bundled gem is installed.
New jekyll site installed in C:/Users/camden/Desktop/ToDelete/myblog.
camden@9XXDCAMDEN  ~\Desktop\ToDelete
>
```

Figure 3.3 The output from scaffolding a new blog with the `jekyll new` command

```
> jekyll serve
Configuration file: C:/Users/camden/Desktop/ToDelete/myblog/_config.yml
            Source: C:/Users/camden/Desktop/ToDelete/myblog
       Destination: C:/Users/camden/Desktop/ToDelete/myblog/_site
 Incremental build: disabled. Enable with --incremental
      Generating...
       Jekyll Feed: Generating feed for posts
                    done in 0.613 seconds.
 Auto-regeneration: enabled for 'C:/Users/camden/Desktop/ToDelete/myblog'
    Server address: http://127.0.0.1:4000/
  Server running... press ctrl-c to stop.
```

Figure 3.4 Example output from running `jekyll serve`

Note that the output from the Jekyll CLI provided a server address and port. If you run `jekyll serve --help` you'll see many different options you can pass to the CLI, including ways to modify the port if you wish. Opening that address in your browser will show you the blog's home page, as seen in figure 3.5.

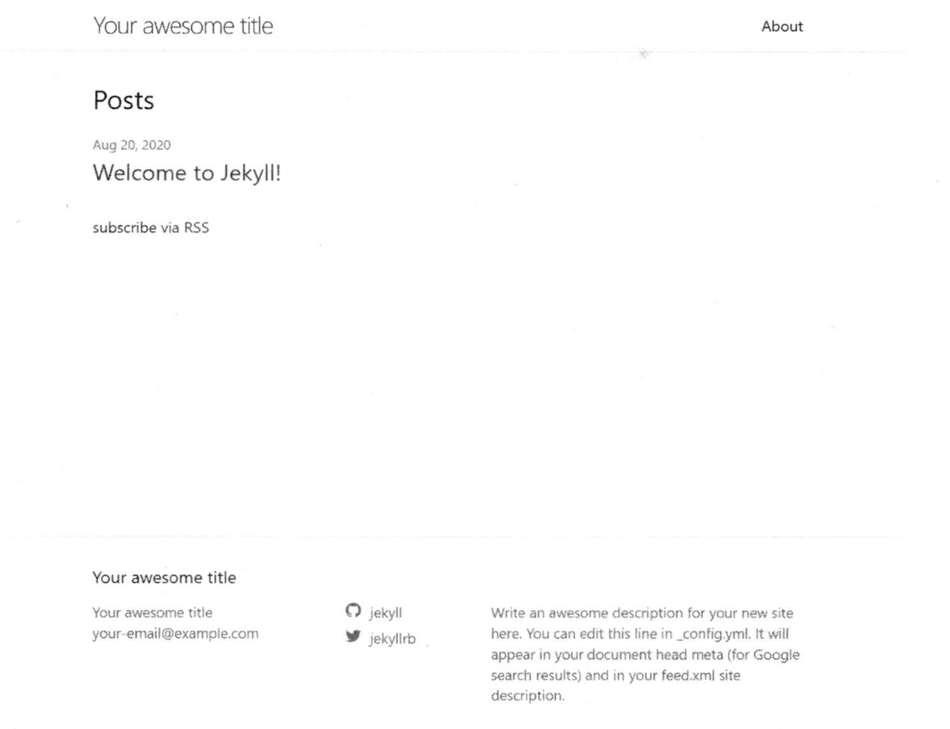

Figure 3.5 The default blog home page

Note that what you are seeing here is a *default* user interface. You are absolutely allowed to change what you see here, and we'll cover this in detail as we continue. To see an example post, click the "Welcome to Jekyll!" link, and you'll see a post, as demonstrated in figure 3.6.

Note that you are not required to scaffold a new site using the `jekyll new` command. You can also simply run `jekyll serve` in a directory with existing HTML files, and Jekyll will run its server from there. The nice thing about the `new` command is that by giving you some common defaults out of the box (a home page, a post, even an RSS feed), you'll save yourself some grunt work.

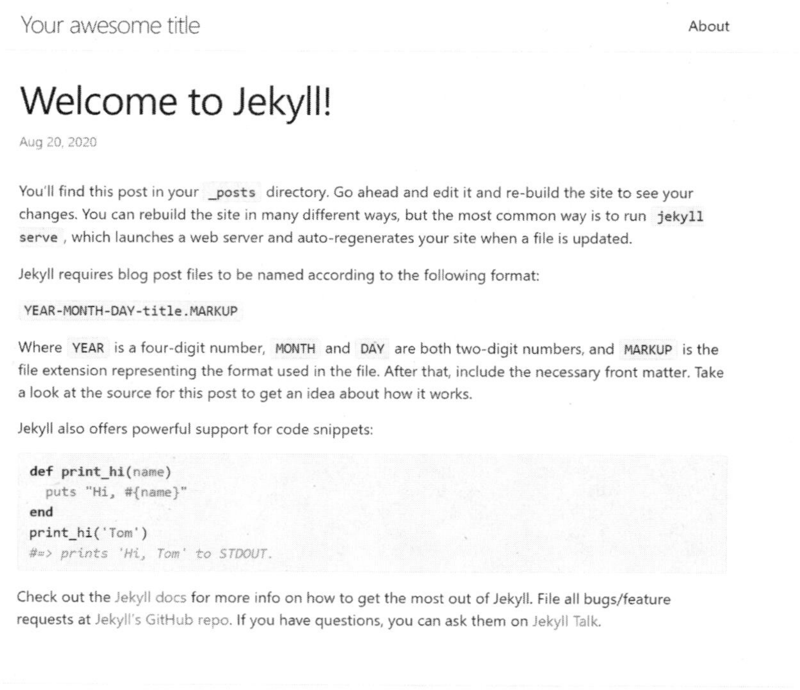

Your awesome title About

Welcome to Jekyll!

Aug 20, 2020

You'll find this post in your `_posts` directory. Go ahead and edit it and re-build the site to see your changes. You can rebuild the site in many different ways, but the most common way is to run `jekyll serve`, which launches a web server and auto-regenerates your site when a file is updated.

Jekyll requires blog post files to be named according to the following format:

`YEAR-MONTH-DAY-title.MARKUP`

Where `YEAR` is a four-digit number, `MONTH` and `DAY` are both two-digit numbers, and `MARKUP` is the file extension representing the format used in the file. After that, include the necessary front matter. Take a look at the source for this post to get an idea about how it works.

Jekyll also offers powerful support for code snippets:

```
def print_hi(name)
  puts "Hi, #{name}"
end
print_hi('Tom')
#=> prints 'Hi, Tom' to STDOUT.
```

Check out the Jekyll docs for more info on how to get the most out of Jekyll. File all bugs/feature requests at Jekyll's GitHub repo. If you have questions, you can ask them on Jekyll Talk.

Figure 3.6 A default welcome blog post for your new Jekyll site

If you look at your scaffolded folder (the directory created by the CLI you just used), you'll see a set of files and folders that define the default blog. While the particulars may change in the future, here's what you can typically expect to see:

- `_posts` is the directory for your blog posts.
- `_site` is where the static version of your Jekyll blog is stored.
- `_config.yml` is a YAML-based configuration file. We'll look into this more later.
- `404.html`, `about.markdown`, and `index.markdown` are files representing a 404 handler (missing file), a simple About page, and the home page for the blog, respectively.
- There are more files Jekyll uses that you can ignore for now.

As you continue learning about Jekyll in this chapter, you'll work with new directories that have special meaning for Jekyll.

One thing you *won't* see are files that handle the layout of the site. Jekyll's default blog uses a "gem-based" theme where the files are stored in your file system outside of the blog itself. When we start customizing the blog later in the chapter we'll work with files *in* the directory instead. You don't have to worry about the location of the gem-based theme, as all its code and assets are copied into the generated folder when you build a Jekyll site.

3.3 Writing a Jekyll post

Your new blog has one simple post, so let's add a new one and see how it reacts. Inside the _posts folder you will see one file named something like 2020-08-22-welcome-to-jekyll.markdown. The reason I say "something like" is that Jekyll will use the current date when generating the file, so the date values in the filename should be values for your current date and time. The format Jekyll expects for blog posts is YEAR-MONTH-DAY-TITLE. Note that both MONTH and DAY should be two-digit numbers, so for numbers less than 10, include the 0 in front.

Create a new file with the current year, month, and date, but with a title of `hello-world`, for example, `2020-08-22-helloworld.markdown`. Jekyll posts require front matter on top, and to start, you can copy the front matter from the first file. Here's a complete file you could use, but remember to update the date for your current time.

Listing 3.1 A new blog post

```
---
layout: post
title:   "Hello World"
date:    2020-08-22 9:00:29 -0500
categories: general
---

Hello World. This is my first *awesome* post!
```

Note that the title is completely up to you, and the contents of the post are arbitrary. However, pay special attention to the time. In my test, the first post Jekyll created had this time stamp: `2020-08-22 08:44:29 -0500`. I wanted my post to be newer, so I used a time a few hours later. However, Jekyll was smart enough to recognize that the date value was in the future and therefore didn't publish the blog post. That's a great feature, but one you could accidentally run into when testing. You can, if you want, also use a date in the past. Jekyll will sort your posts based on this value automatically. Finally, why do we have date-based filenames if the front matter has a date value as well? First, having the filenames be date-based will give you a visual cue when looking at your directory in terms of what content was written when. Second, if you leave off the date value in the front matter, Jekyll will use the date from the filename for the post. When making posts, you can decide if you care about the time value of posts, as Jekyll will be fine with either way of specifying the date.

> **NOTE** Jekyll has a command-line flag to show future posts when working locally. You can enable the feature by running `jekyll serve --future`.

The value for `layout` specifies how to render the post. Remember that the default theme "hides" the details for us, but later when we start building our own layout we will be able to customize this. Finally, `categories` is—you guessed it—a way to categorize posts. Many blogs will have category pages that let you browse posts from one category. Jekyll supports this as well, but not out of the box with the default theme.

Once you've saved your file, go back to your browser and reload. You should see the new post on the home page, as demonstrated in figure 3.7.

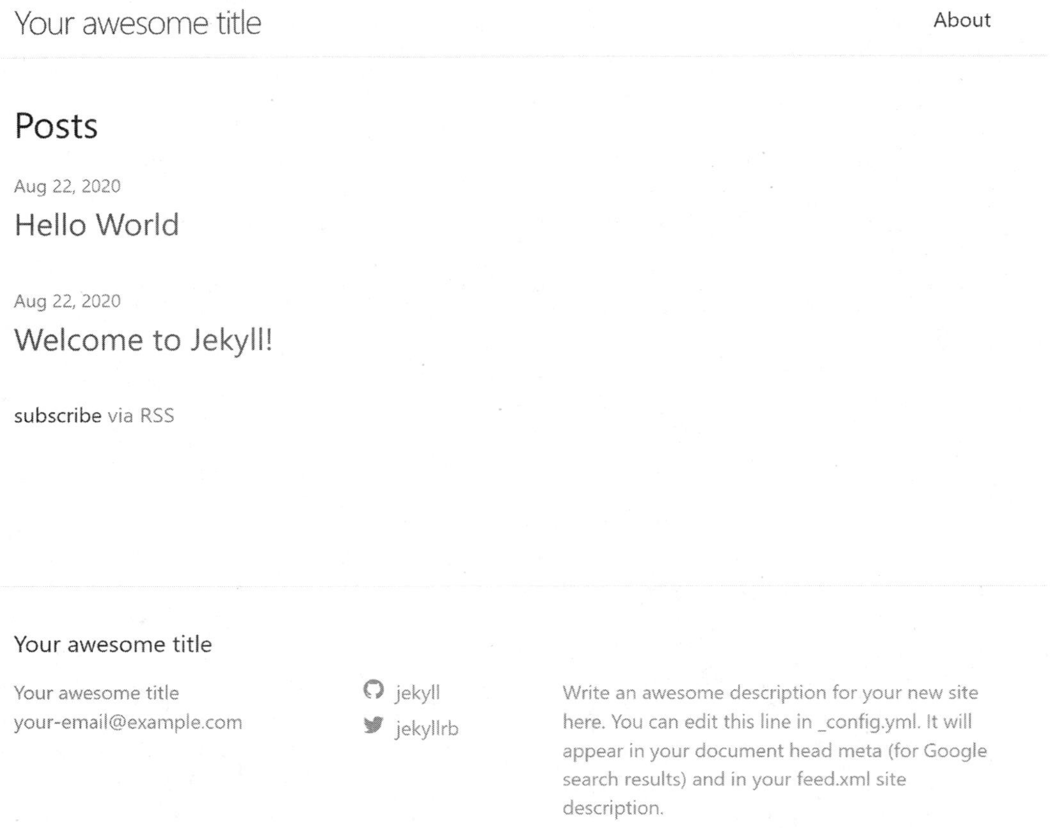

Your awesome title
About
Posts
Aug 22, 2020
Hello World
Aug 22, 2020
Welcome to Jekyll!
subscribe via RSS
Your awesome title
Your awesome title
your-email@example.com
jekyll
jekyllrb
Write an awesome description for your new site here. You can edit this line in _config.yml. It will appear in your document head meta (for Google search results) and in your feed.xml site description.

Figure 3.7 After adding a new post, it will show up on the home page.

If you click on the post, you'll see it rendered. Figure 3.8 shows how Jekyll rendered the Markdown into HTML and automatically applied the theme's layout.

Make note of the URL for your new post. Again, remember that the dates will differ; it should look something like this: http://localhost:4000/general/2020/08/22/helloworld.html. Notice how Jekyll parsed your initial filename and created a path based on it while also including the category of the post as well. Personally, I don't care for this format, but luckily Jekyll lets you customize this format.

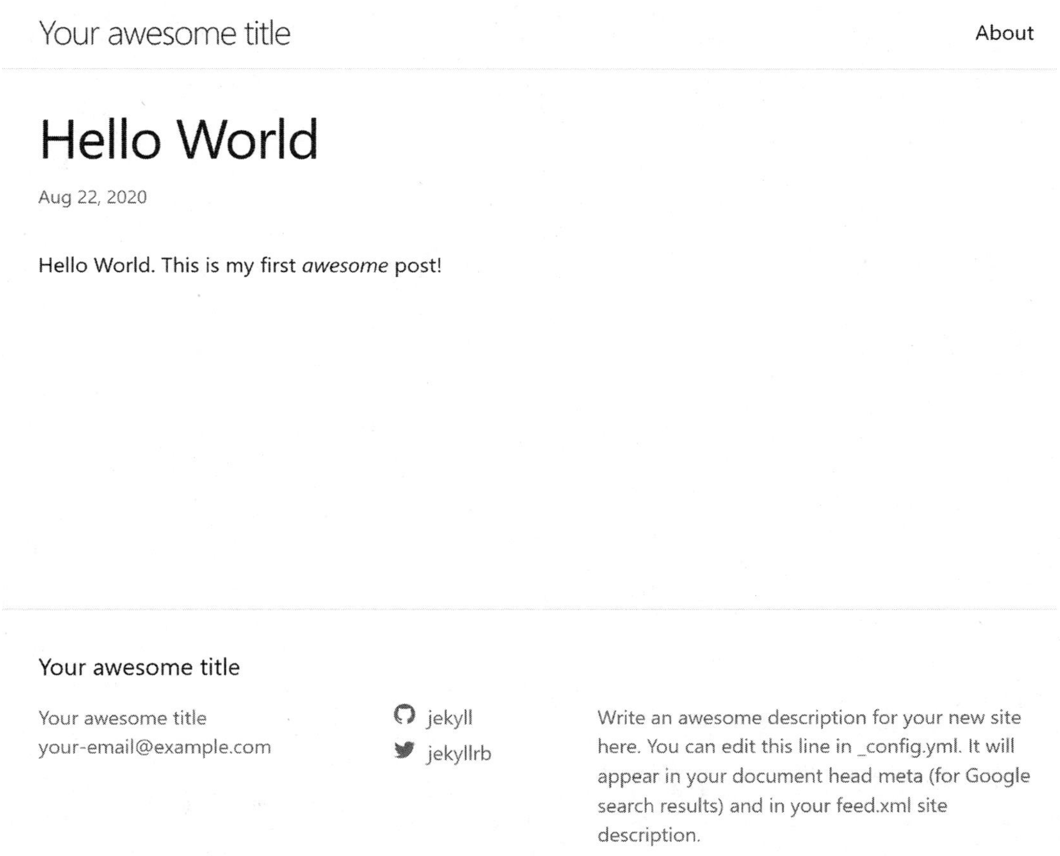

Your awesome title About

Hello World

Aug 22, 2020

Hello World. This is my first *awesome* post!

Your awesome title

Your awesome title
your-email@example.com

Write an awesome description for your new site here. You can edit this line in _config.yml. It will appear in your document head meta (for Google search results) and in your feed.xml site description.

Figure 3.8 Your new blog post rendered by Jekyll

3.3.1 *A liquid refresher*

Jekyll uses the Liquid template language to allow for dynamic code while generating static pages. In the previous chapter, we covered the basics of using Liquid in section 2.1.2, but in case you skipped that chapter, let's cover some quick basics.

First, Liquid code uses two kinds of markup. To output a simple value, you would use `{{ variable }}`, where `variable` is the value you want to output. To execute a line of code, a slightly different form is used: `{% %}`. This command will assign a value to a variable: `{% assign foo = "cat" %}`. Liquid has support for both conditionals and looping. Next, we modify the blog post you just created to include the same sample code from the previous chapter.

Listing 3.2 Blog post with Liquid commands

```
---
layout: post
title:  "Hello World"
date:   2020-08-22 9:00:29 -0500
categories: general
---

Hello World. This is my first *awesome* post!

{% assign name = "ray" %}

<p>
Hello, {{ name }}!
</p>

{% assign cool = true %}

{% if cool %}
<p>
Yes, you are cool.
</p>
{% endif %}

{% assign cats = "Fluffy,Muffy,Duffy" | split: ',' %}
<ul>
{% for cat in cats %}
    <li>{{ cat }}</li>
{% endfor %}
</ul>
```

If you save this and reload, you'll see the result in figure 3.9; feel free to modify the values and play around to see the changes in action.

Jekyll will parse your Liquid in blog posts and in regular pages on your site. This feature also works if you use HTML files instead of Markdown.

Hello World

Aug 22, 2020

Hello World. This is my first *awesome* post!

Hello, ray!

Yes, you are cool.

- Fluffy
- Muffy
- Duffy

Figure 3.9 A Jekyll blog post with Liquid code adding dynamic aspects to the static page

3.4 Working with layouts and includes

So far, the particulars of how content is displayed have been abstracted. We explained that Jekyll is using a default theme and that the theme files themselves are stored "elsewhere," but how do they actually work, and how would you modify them? Let's start by creating a new blog using the CLI again: `jekyll new blog1`. (The listings in this section can all be found in the GitHub repository for this book in the chapter3/blog1 folder.) When it's done scaffolding the new site, start the server with `jekyll serve`. To begin working with layouts, create a new directory named _layouts. By default, Jekyll will look in this folder for layout files. (You can customize the name of this folder, if you wish.) If you open the index.html file scaffolded by the CLI you'll notice that it specifies a layout named `home` (note that I've removed the comments):

```
---
layout: home
---
```

Let's change this value to `default`. The Jekyll docs suggest using a layout named default and then extending it when you have different layouts. You'll see an example of this in a moment. But for now, change the value in the front matter to default:

```
---
layout: default
---
```

Now let's create a layout file. Listing 3.3 demonstrates a very minimal but custom layout. Since we specified default as the layout, the filename must match, so name this file default.html.

Listing 3.3 The new layout file (_layouts/default.html)

```
---
title: Default Title
---

<html>
<head>
<title>{{ page.title }}</title>     <- The page.title variable will come from pages using the layout.
</head>

<body>

{{ content }}     <- The content variable is the page content itself that uses the layout.

</body>
</html>
```

This template is about as simple as you can get. There's no CSS or layout at all really, just a few HTML tags. There are two variables being used here. The first, `page.title`, will be replaced with the title of the page using the layout. The second, `content`, will be replaced by the content of the page itself. If you reload your blog now you'll see … nothing! Why? The logic you saw earlier (creating a list of blog posts) was done by the default theme. With that removed, your home page is unfortunately empty. Let's fix that by adding the posts back in.

Listing 3.4 The new home page (/index.markdown)

```
---
layout: default
title: My Blog
```

```
---

<h1>Posts</h1>

<ul>
{% for post in site.posts %}
<li><a href="{{ post.url }}">{{ post.title }}</a>, written
{{ post.date}}</li>
{% endfor %}
</ul>
```

The first thing we've added here is a title in the front matter. The layout file will notice this and use it in the page's header. Next, we use Liquid to iterate over a variable, `site.posts`. Jekyll provides this variable automatically, based on the files within your _posts directory. Each post has a `url` and `title` value that can be used to render links and titles for the posts. Finally, you can also display the date from the post. Using an unordered list to display posts isn't terribly pretty, but it gets the job done, as shown in figure 3.10.

Posts

- Welcome to Jekyll!, written 2020-08-22 10:05:52 -0500

Figure 3.10 Our new, boring, home page

3.4.1 Layout inheritance

Jekyll supports the idea of *layout inheritance*, which is just a fancy way of saying one layout that wraps another layout. Let's build a simple example of this by customizing how blog posts are displayed. If you open the default blog post that the CLI scaffolded, you'll see it's already specifying a unique layout, `post`. In your _layouts folder, create a new file named post.html.

Listing 3.5 The post layout (_layouts/post.html)

```
---
layout: default
---

<h1>Blog Post: {{ page.title }}</h1>

{{ content }}
```

Again, this is a rather boring layout, but make note of how it specifies a layout itself. This means that the post layout will display its contents and then pass them on to the next layout. In this case, our post layout simply includes an h1 tag on top with the title of the post. Figure 3.11 demonstrates our new post layout.

Blog Post: Welcome to Jekyll!

You'll find this post in your `_posts` directory. Go ahead and edit it and re-build the site to see your changes. You can rebuild the site in many different ways, but the most common way is to run `jekyll serve`, which launches a web server and auto-regenerates your site when a file is updated.

Jekyll requires blog post files to be named according to the following format:

`YEAR-MONTH-DAY-title.MARKUP`

Where `YEAR` is a four-digit number, `MONTH` and `DAY` are both two-digit numbers, and `MARKUP` is the file extension representing the format used in the file. After that, include the necessary front matter. Take a look at the source for this post to get an idea about how it works.

Jekyll also offers powerful support for code snippets:

```
def print_hi(name)
  puts "Hi, #{name}"
end
print_hi('Tom')
#=> prints 'Hi, Tom' to STDOUT.
```

Check out the Jekyll docs for more info on how to get the most out of Jekyll. File all bugs/feature requests at Jekyll's GitHub repo. If you have questions, you can ask them on Jekyll Talk.

Figure 3.11 A post using the new customized layout

3.4.2 Using includes

Now let's look at includes. Unlike layouts, which wrap content, an *include* is simply content that is inserted into a file. The format for including files is this Liquid code: `{% include file %}`. Jekyll will look for these files in a new folder, _includes. Let's test this by adding a copyright notice to our site. First, make the new _includes directory, and then use the following code for the contents.

Listing 3.6 The copyright file (_includes/copyright.html)

```
<p>
Copyright {{ "now" | date:"%Y" }}
</p>
```

The Liquid code used here displays the current year based on when the site was last built. The filter, `date`, is passed a formatting string, which in this case just asks for the year. The value `"now"` represents the current time. Remember that static sites are static, so your statically generated website won't magically show a new value a second after the ball drops for New Year's Eve in New York City, but once you rebuild the site, every template using this include will have the right value. To use this value, modify the default layout file.

Listing 3.7 Updated default layout file (_layouts/default.html)

```
---
title: Default title
---

<html>
<head>
```

```
<title>{{ page.title }}</title>
</head>

<body>

{{ content }}

{% include copyright.html %}
</body>
</html>
```

The include here specifies that the content will come right before the closing body tag.

If you reload your blog now, you'll see a copyright notice on every page, as seen in figure 3.12.

Posts

- Welcome to Jekyll!, written 2020-08-22 10:05:52 -0500

Copyright 2020

Figure 3.12 The all-important copyright notice, now displayed on our site

3.5 Creating additional files

When you used the Jekyll CLI to scaffold a blog, it created an additional file named about.html. When working on pages that aren't blog posts, you can simply create any HTML or Markdown file in the root directory of your site, and Jekyll will include it in the final build. While that's pretty much all there is to it, let's add a new page to the blog to ensure the process sinks in. Listing 3.8 displays the contents of a Contact Us page that is common on websites.

Listing 3.8 The new contact page (/contact.md)

```
---
layout: default
title: Contact Us
---

## Contact Us

Please send us an email at some random email address
that never gets checked. Or you call us at 555-555, but we
probably won't answer. We don't do faxes because it's 2020.
```

First, notice that while Jekyll used `.markdown` you can also use `.md` as an extension, if you prefer. Once saved, Jekyll automatically makes the page available at /contact .html. If you open the About page (about.markdown) that Jekyll generated, you'll notice it uses a permalink value in the front matter to specify another path, /about/. If you prefer that style of naming your file, you can add it to the front matter of the contact page as well.

TIP If you encounter issues with Jekyll, errors, or warnings about Gemfiles, a common solution is to run a bundle update or use `bundle exec jekyll serve`.

3.6 Working with data

You've seen how to work with blog posts and simple pages in Jekyll, but Jekyll also allows you to work with data. This data could be anything: perhaps a list of authors for a blog and their contact information. Your site pages can read this data and then display it in templates.

To start, you'll need to create a _data folder. Inside this folder you can create files in either JSON, CSV, TSV, or YAML formats. The name of the file will control how it's available later in templates. Let's consider a simple example. (The listings in this section can all be found in the GitHub repository for this book in the chapter3/blog2 folder.)

Listing 3.9 Our blog authors (/_data/authors.json)

```
[
    {
    "name":"Raymond Camden",
    "website":"https://www.raymondcamden.com",
    "twitter":"raymondcamden"
    },
    {
    "name":"Brian Rinaldi",
    "website":"https://remotesynthesis.com/",
    "twitter":"remotesynth"
    }
]
```

The data here is arbitrary and can be anything that makes sense for your site. To access this data, Jekyll will make it available in templates as `site.data.authors`. You can also use subdirectories in your data folder, and if you do, the name of the subdirectory will be added to the variable structure. If our authors.json file was in a subdirectory named people, the new variable to access that data would be `site.data.people.authors`. Now let's edit the blog's home page to list our authors.

Listing 3.10 An updated home page with authors (/index.markdown)

```
---
layout: default
title: My Blog
---

<h1>Posts</h1>

<ul>
{% for post in site.posts %}
<li><a href="{{ post.url }}">{{ post.title }}</a>, written
{{ post.date}}</li>
{% endfor %}
```

```
</ul>

<h2>Our Authors</h2>

<ul>
{% for author in site.data.authors %}
<li>
    <a href="{{site.website}}">{{ author.name }}</a> -
    <a href="https://twitter.com/{{author.twitter}}">@{{author.twitter}}</a>
</li>
{% endfor %}
</ul>
```

Looping over the array of authors

The changes begin after the list of posts. You can see the Liquid tag being used to iterate over the site data. For each author, we display their name with a link to their website as well as linking to their Twitter profile. Figure 3.13 shows how this is displayed.

Posts

- Welcome to Jekyll!, written 2020-08-22 10:05:52 -0500

Our Authors

- Raymond Camden - @raymondcamden
- Brian Rinaldi - @remotesynth

Copyright 2020

Figure 3.13 The authors displayed are driven by a JSON file.

3.7 Configuring your Jekyll blog

As mentioned earlier in the chapter, the Jekyll CLI creates a file that configures how Jekyll works. This file is named _config.yml, and as you can tell by the extension, it uses YAML for formatting.

> **NOTE** YAML is a "simple" text-based format but can sometimes be a bit confusing. To learn more about the syntax, see the official website at https://yaml.org/.

There are numerous values you can configure, so for a complete list of settings, be sure to see the Jekyll docs at https://jekyllrb.com/docs/configuration/options/, but here are a few values you may want to know right away:

- `title`—You can specify a site-wide title for a site by specifying a value in your config file. This will be available in templates as `site.title`. Note that any and all values specified in the config file are available as site variables.
- `exclude` and `include`—Specifies files and folders to either ignore or to be forcibly included. Dotfiles are not included by default, so this lets you include them

when specified. The exclude feature is good when working with very large sites, as it lets you ignore sets of files in development but not production.

- `Permalink`—Earlier in the chapter, we demonstrated how Jekyll blog posts will include the category in the URL. This is because of the default value for permalinks: `/:categories/:year/:month/:day/:title:output_ext`. To "fix" this, you can specify a permalink value that removes the `:categories:` portion. There are multiple tokens you can use in your permalink string, and you can find them documented at https://jekyllrb.com/docs/permalinks/#placeholders.

3.8 Generating your site

While working with the Jekyll server running, you may have noticed it outputting the results to a directory named _site. This is the default output directory for your Jekyll builds. You can generate a build at the command line manually by running `jekyll build`. You'll get a report of the operation, as seen in figure 3.14.

```
PS C:\projects\the-jamstack-book\chapter3\blog2> jekyll build
Configuration file: C:/projects/the-jamstack-book/chapter3/blog2/_config.yml
            Source: C:/projects/the-jamstack-book/chapter3/blog2
       Destination: C:/projects/the-jamstack-book/chapter3/blog2/_site
 Incremental build: disabled. Enable with --incremental
      Generating...
       Jekyll Feed: Generating feed for posts
                    done in 0.624 seconds.
 Auto-regeneration: disabled. Use --watch to enable.
PS C:\projects\the-jamstack-book\chapter3\blog2> |
```

Figure 3.14 The output from requesting that Jekyll build your site

You can configure the output directory by using the `--destination` flag, for example, `jekyll build --destination output`. Figure 3.15 demonstrates what this looks like for the blog from the previous section.

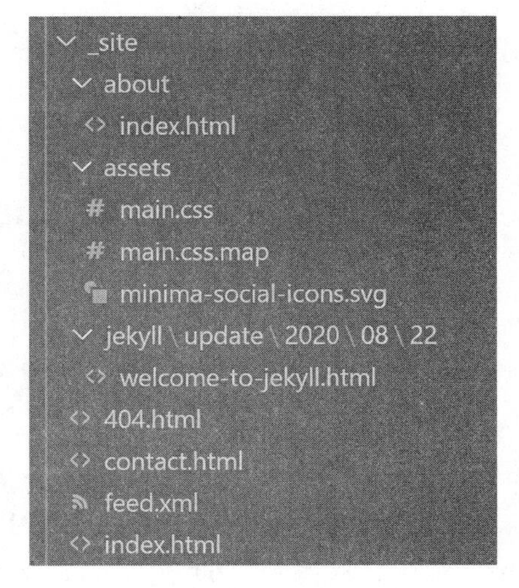

Figure 3.15 HTML output generated by the Jekyll build command

3.9 Building a Jekyll blog

Since Jekyll can scaffold a blog out of the box, technically there isn't much we need to discuss here, but it would be nice to demonstrate the process someone goes through in modifying the blog with a different look and feel. You saw earlier how to build a custom layout, but there are many themes out there that give you a complete look and feel with little to no work on your part.

A great example of this is the "Clean Blog Jekyll" theme from Start Bootstrap (https://startbootstrap.com/themes/clean-blog-jekyll/). This is a free blogging theme that you can use to start your own projects, and, of course, you can use the theme as a starting point, modifying small things here and there to get it just right.

You can find install directions at the theme's GitHub repository, both for adding the theme to a blog scaffolded by the CLI as well as for using a copy of the repository itself. That option is somewhat easier, so download a zip of the repository and extract it.

Once extracted, change to the directory in your terminal and run Jekyll with `bundle exec jekyll serve`. This will fire up the blog locally, as seen in figure 3.16.

```
PS C:\Users\ray\Downloads\startbootstrap-clean-blog-jekyll-master> bundle exec jekyll serve
Configuration file: C:/Users/ray/Downloads/startbootstrap-clean-blog-jekyll-master/_config.yml
            Source: C:/Users/ray/Downloads/startbootstrap-clean-blog-jekyll-master
       Destination: C:/Users/ray/Downloads/startbootstrap-clean-blog-jekyll-master/_site
 Incremental build: disabled. Enable with --incremental
      Generating...
       Jekyll Feed: Generating feed for posts
                    done in 53.147 seconds.
 Auto-regeneration: enabled for 'C:/Users/ray/Downloads/startbootstrap-clean-blog-jekyll-master'
    Server address: http://127.0.0.1:4000/startbootstrap-clean-blog-jekyll/
  Server running... press ctrl-c to stop.
```

Figure 3.16 Output from running the downloaded Jekyll theme

Pay special attention to the server address. Unlike earlier examples that started in the root of a web server, this one starts in a subdirectory. We'll fix that in a moment, but to be sure everything is okay, open the URL in your browser, and you should see the theme in use with some default blog posts (figure 3.17).

Man must explore, and this is exploration at its greatest

Figure 3.17 The Clean Blog running locally

Now it's time to start customizing this blog and making it your own. Use `ctrl+c` to shut down the current running server and open _config.yml. The following code shows the top half of the config file after changing it to make it personal for the author.

Listing 3.11 New configuration values (/_config.yml)

```
title:              My Blog
email:              raymondcamden@gmail.com
description:        My New Blog!
author:             Raymond Camden
#baseurl:             "/startbootstrap-clean-blog-jekyll"
#url:                 "https://startbootstrap.github.io"

# Social Profiles
twitter_username:   raymondcamden
github_username:    cfjedimaster
facebook_username:
instagram_username:
linkedin_username:
```

The first four values were changed to match my information, but you should absolutely type something else here instead. I commented out both `baseurl` and `url` since they aren't needed for our test, and we can accept defaults Jekyll will use. Under `Social Profiles`, I specified a few of the values I felt like sharing. The blog theme is going to look for these values and render itself differently based on them. Save your changes and run the server again. Now when you view it (make note of the new URL), you'll see custom values, as seen in figure 3.18.

Figure 3.18 The blog now has a custom title and other values. Yours will look different.

That was easy, right? What you're seeing here is the theme exposing values in the configuration that are then picked up and used in the display. Let's make another change. While that header picture is pretty, we can pick another. There is a wonderful website, Unsplash.com, that provides beautiful stock photography free of charge. In figure 3.19, you can see a picture I selected (https://unsplash.com/photos/V705bwrTnQI) by Annie Spratt (https://unsplash.com/@anniespratt).

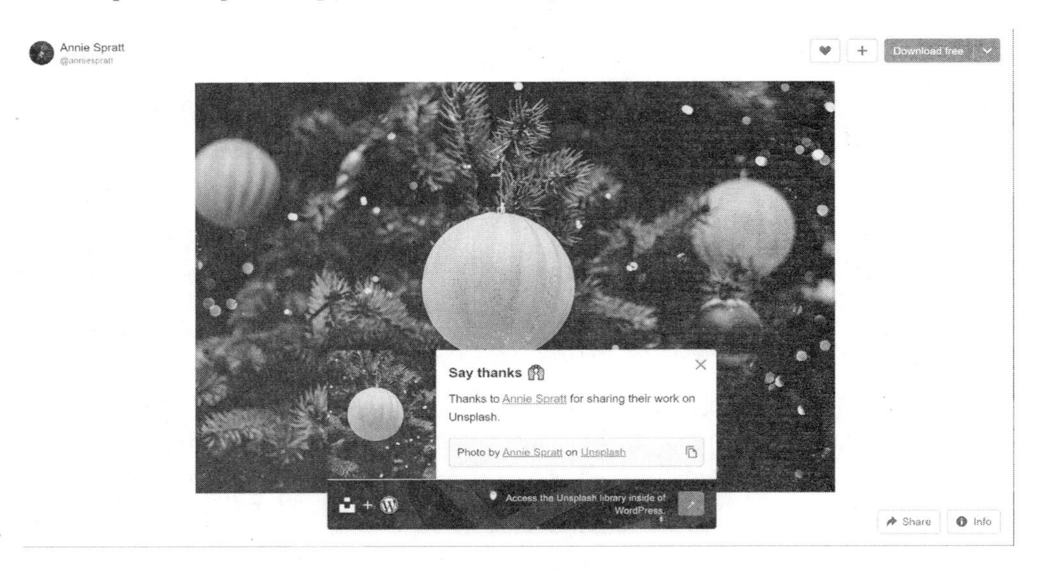

Figure 3.19 Free stock photography from the Unsplash.com site

After downloading the image, rename it to xmastree.jpg and save it in the blog's img folder. The default images from Unsplash can be rather large, so the copy you'll find in the book's repository has been resized. The blog's home page can be found in the index.html file. Open it and you'll see it's rather short.

Listing 3.12 The blog home page (/index.html)

```
---
layout: home
background: '/img/bg-index.jpg'
---
```

In this case, the blog's theme handles the logic of getting blog posts and rendering them, so to update the large image on top, you can simply edit the background value.

Listing 3.13 Update blog home page (/index.html)

```
---
layout: home
background: '/img/xmastree.jpg'
---
```

Save this change and reload the blog (figure 3.20).

Figure 3.20 Our new blog header (and again, you can pick anything!)

The last thing we'll do is a bit of cleanup. The blog comes with six blog posts. Normally when working with a new blog theme, I'll delete all but one I keep as a template for further blog posts. Delete all but one, for example, 2020-01-26-dinosaurs.html. Open it, remove the sample content, and change the title and date values.

Listing 3.14 The updated blog post

```
---
layout: post
title: "Welcome to my blog"
subtitle: "I'm so excited!"
date: 2020-08-24 12:00:00 -0400
background: '/img/posts/01.jpg'
---

<p>
This is my cool blog!
</p>
```

As stated many times before, feel free to use your own title, subtitle, and content. The date should also be something more appropriate for when you are reading this book. You can also change the image, if you wish, to something you like. Lastly, while not required, you should rename the file. Remember that the date in the front matter will take priority over the date in the filename. But for housekeeping in general, the filename should better represent the content, for example, 2020-08-24-welcome.html. After saving your changes, you can reload the blog and click on your remaining blog title to see it in all its glory (figure 3.21).

My Blog
HOME
ABOUT
POSTS
CONTACT
Welcome to my blog
I'm so excited!
Posted by Raymond Camden on August 24, 2020 · 1 min read
This is my cool blog!

Figure 3.21 The newly edited blog entry

To go forward with this blog, you would simply continue adding new blog posts, and the theme will take care of adding them for you automatically.

3.10 Going further with Jekyll

While we demonstrated some of the cool features of Jekyll, there's a lot we didn't have space to cover. Some of these topics include the following:

- While Jekyll is focused on building blogs, it has a feature called Collections (https://jekyllrb.com/docs/collections/) that allows you to define your own lists/sets of data in whatever form you desire. This can be useful for documentation sites or other related content.
- Jekyll also has built-in support for pagination (https://jekyllrb.com/docs/pagination/), providing a quick way to set up how many items would be on a page as well as creating variables your code can use for creating links to previous and forthcoming pages of content.
- For cases where Jekyll doesn't do enough, it has a plug-in system (https://jekyllrb.com/docs/plugins/) that lets you extend its functionality. This will require you to write Ruby, though.
- Finally, don't forget that GitHub has native support for running Jekyll-based sites for repositories with content (like documentation).

For more information, including links to support, see the Jekyll Community page at https://jekyllrb.com/docs/community/.

In the next chapter, you'll see how to build a site focused on providing documentation. You'll look at the Hugo static site generator, easily one of the fastest ones available (which makes it great for large sites).

Summary

- Jekyll is static site generator with a special focus on building blogs.
- Jekyll uses Liquid as a template language and lets you write other pages in Markdown or HTML.
- Jekyll uses Ruby for installation, so be prepared for (possibly) a harder time under Windows.
- Jekyll supports global data files written in JSON, CSV, TSV, and YAML.

4 Building a documentation site

This chapter covers

- Understanding the needs of a typical documentation site
- Choosing a headless CMS for managing documentation
- Choosing a static site generator for a documentation site
- Installing and configuring Hugo
- Setting up Netlify CMS open editing and modeling
- Configuring Netlify and GitHub for user authentication
- Editing content in Netlify CMS

The Jamstack has always excelled at content-focused sites, even from the early days of static site generators. Static HTML and CSS is perfect for displaying content quickly and efficiently; thus, content sites lend themselves to pre-rendering using Jamstack tools. This is why documentation sites have been one of the most obvious use cases for the Jamstack.

Documentation sites have always generally had additional advantages to using the Jamstack:

- The ability to version file-based content easily via source control
- A means of accepting contributions and corrections via processes like a GitHub pull request
- The fact that, in many cases, the authors were technically adept with these sorts of development tools

The biggest disadvantage of choosing the Jamstack for a documentation site had typically been that the tools for editing content were not advanced enough to meet the needs of content authors and editors. However, the tools, services, and libraries available for documentation sites (or content-focused sites in general) using the Jamstack have improved tremendously in recent years. The benefits still apply, but modern Jamstack tools make site content easier to edit and contributions easier to accept—even from folks who may be unfamiliar with code. In this chapter, we'll explore the options available to you for developing documentation sites using the Jamstack and walk through how to build a documentation site.

4.1 Requirements of a documentation site

There's obviously no single kind of documentation site. For example, there is technical documentation, like software documentation or API documentation, and end-user documentation, like user manuals. The requirements for each of these may differ, but they also share some commonalities:

- Documentation sites tend to have multiple, and often numerous, contributors. In the case of project or policy documentation, contributors might exclusively be company employees. However, in a software world increasingly dominated by open source, many documentation sites often have a large number of external contributors.
- Contributors may have varying degrees of technical expertise when it comes to editing the site's content.
- Documentation sites are designed to be optimized for fast and easy access to information, favoring simple and straightforward features and design.
- The layout and design are geared toward readability over style.
- Outside of things like comments or runnable examples, many documentation sites do not include complex, dynamic functionality.
- Many documentation sites have a lot of content, but that content can change relatively infrequently. A typical site may receive periodic major updates with occasional minor ones in between.

4.1.1 The example site requirements

We're going to build technical documentation for an esoteric programming language called LOLCODE (http://www.lolcode.org/) (see figure 4.1). LOLCODE is intended to be a humorous take on a programming language that is based on lolspeak, a form of grammatically incorrect language associated with internet memes of cats. The docs are based on the LOLCODE specification (http://mng.bz/RE7D), written by Justin Meza.

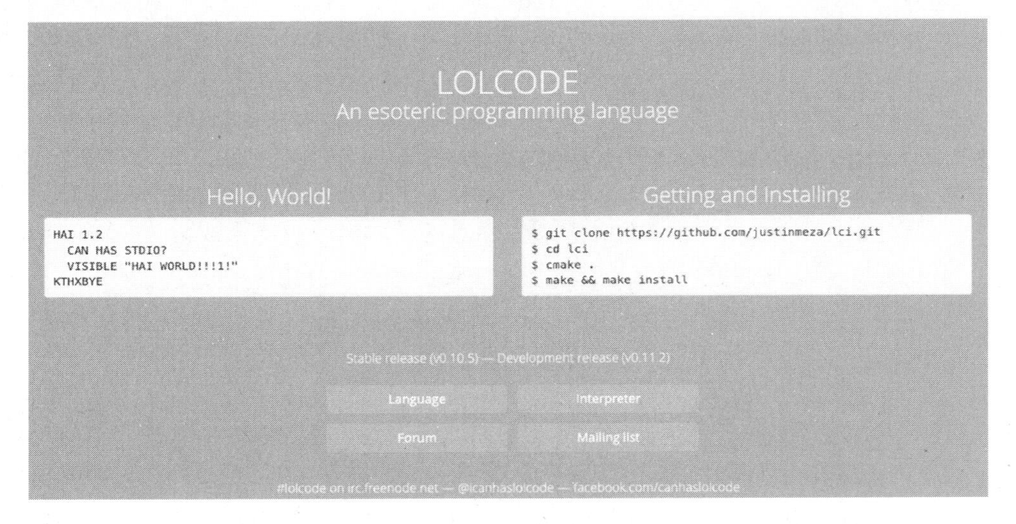

Figure 4.1 The lolcode.org site showing the LOLCODE language, installation details, and links to the specifications that we'll use for our example documentation site

As the adoption of LOLCODE inevitably expands, I expect this site to get rather large, so optimizing for build speed will be a priority. Since this is an open source language, I'd also like to enable easy contributions from third-party authors. Though I anticipate that most contributors will have a high degree of technical expertise, they may not be experts in building Jamstack sites. Nonetheless, I want to encourage contributions by enabling them to contribute without requiring that they go through the Git workflow of manually forking, installing locally, updating, and then creating a pull request.

Now that we understand the needs of our docs site, let's look at the tools we have available to meet these requirements.

4.2 Choosing the right tools

Our requirements aren't particularly complex. We need the ability to generate potentially many content pages for an extensive documentation, and we need the ability for third-party content contributors to edit content without requiring a deep technical knowledge of how the site is built. To accomplish this, we're going to need the right static site generator and a headless content management system (CMS). The CMS will provide the editing interface that will allow content contributors to more easily write and edit content on the site.

4.2.1 What is a headless CMS?

We've talked a lot about static site generators but not about headless CMSs. *Headless CMSs* are a relatively new concept. The name derives from the idea that they decouple the backend (i.e., the actual content editing and management tools that the CMS provides) from the "head" (i.e., the frontend of the application, in this case, a website).

Traditional CMSs were created almost exclusively to manage web page content. Because of this, the management of the content was tied to its display. For example, in a typical WordPress site, the backend content management is provided by WordPress, but the frontend website is also built in (i.e., tightly coupled with) WordPress.

This tight coupling means that content is not reusable. A headline on the home page might also appear on a landing page, but updating one instance will not update the other. Since the content is intended for the web, it cannot easily be reused in things like a mobile app.

A traditional CMS also isn't designed for the Jamstack. Pages on the frontend of a traditional CMS are server-rendered and unable to take advantage of the improved speed and security the Jamstack architecture offers.

Headless CMSs solve these problems by providing the backend content editing and management tools, untethered from a frontend site. There is a fast-growing list of headless CMS options available, but they fall under two different categories that determine how the frontend site accesses the content:

- *API-based headless CMS*—Your content is stored by the CMS provider and is accessible by your website, mobile app, or other application via an API. Since content in an API-based headless CMS is not tied to physical files, they are able to easily handle reuse of content objects and more easily manage complex relationships between content objects or even embedding content objects within content blocks.
- *Git-based headless CMS*—These CMSs do not store your content. Instead, content is maintained in a Git repository, often as Markdown for long-form content and YAML or JSON for data. The CMS is essentially a layer of tooling for managing the content via a web interface that is easy to understand for content editors who may be uncomfortable manually editing the file-based content.

NOTE For a more in-depth look at the pros and cons of Git-based versus API-based headless CMSs, see this detailed post by Bejamas: http://mng.bz/2j19.

One of the benefits of a Git-based CMS for our example use case is that it still allows a Git-based editing workflow and version histories that track changes, which can be publicly accessible via GitHub or other Git project hosting providers. This can be ideal for a technical documentation project, particularly for an open source project, as in our case. Therefore, our example project will use a Git-based solution.

4.2.2 Headless CMS options

Now that we've determined a Git-based solution is our choice, let's look at some the options available.

FORESTRY

Forestry (forestry.io) is a commercial Git-based CMS solution. It comes with built-in support for all of the most popular static site generators and integrates with most of the major static hosting providers (figure 4.2). At the time of this writing, it offers a free account that supports up to five editors, though free sites are automatically archived after three months of inactivity.

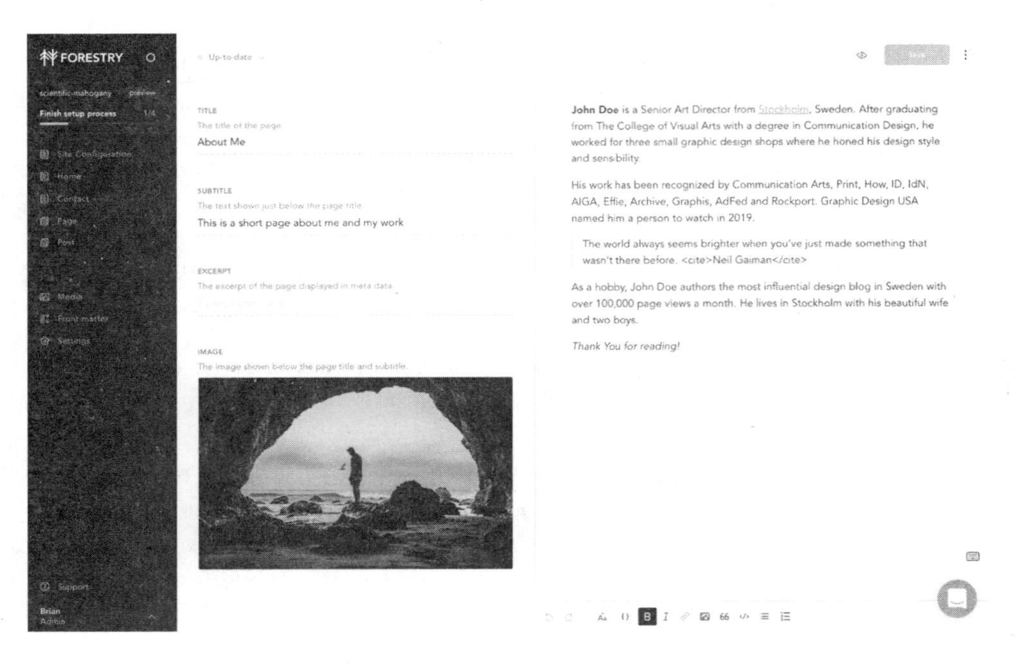

Figure 4.2 Editing a content page using Forestry's what-you-see-is-what-you-get (WYSIWYG) page editor

PUBLII

Publii (https://getpublii.com/), shown in figure 4.3, is different from the other options in that it is an installable, open source desktop application rather than a web-based interface. It offers several options for editing content, including a WYSIWYG

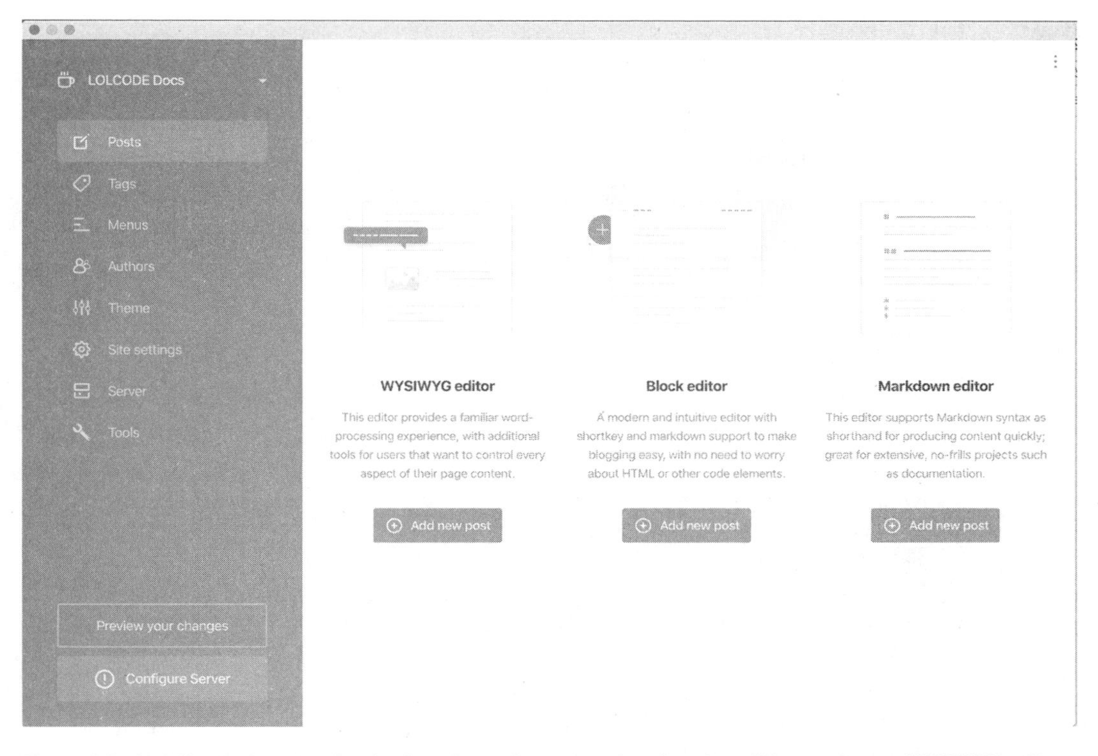

Figure 4.3 Publii's desktop application interface offers several options for editing content: a WYSIWYG editor, a block editor similar to the new "Gutenberg" interface on WordPress, and straight Markdown source editing.

editor, a block editor similar to the new "Gutenberg" interface on WordPress, and a straight Markdown editor. Publii doesn't just provide the editing tools like other Git-based CMSs, but also serves as the static site generator, which supports a wide variety of hosting options.

PROSE

Prose (http://prose.io/), shown in figure 4.4, is also a unique option. It is a completely free tool that hooks into your GitHub account, giving you access to a web-based editor for any file in any repository connected to your account. While code and data files can be edited using the Prose editor, its focus is on offering a better editing experience for Markdown content with metadata (typically called front matter). It does provide a simple Markdown preview but not a true WYSIWYG experience.

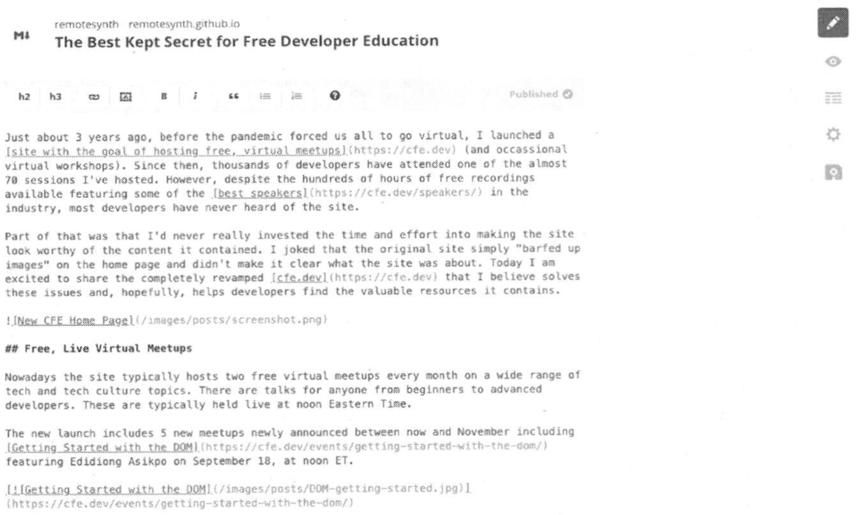

```
remotesynth remotesynth.github.io
The Best Kept Secret for Free Developer Education

Just about 3 years ago, before the pandemic forced us all to go virtual, I launched a
[site with the goal of hosting free, virtual meetups](https://cfe.dev) (and occassional
virtual workshops). Since then, thousands of developers have attended one of the almost
70 sessions I've hosted. However, despite the hundreds of hours of free recordings
available featuring some of the [best speakers](https://cfe.dev/speakers/) in the
industry, most developers have never heard of the site.

Part of that was that I'd never really invested the time and effort into making the site
look worthy of the content it contained. I joked that the original site simply "barfed up
images" on the home page and didn't make it clear what the site was about. Today I am
excited to share the completely revamped [cfe.dev](https://cfe.dev) that I believe solves
these issues and, hopefully, helps developers find the valuable resources it contains.

![New CFE Home Page](/images/posts/screenshot.png)

## Free, Live Virtual Meetups

Nowadays the site typically hosts two free virtual meetups every month on a wide range of
tech and tech culture topics. There are talks for anyone from beginners to advanced
developers. These are typically held live at noon Eastern Time.

The new launch includes 5 new meetups newly announced between now and November including
[Getting Started with the DOM](https://cfe.dev/events/getting-started-with-the-dom/)
featuring Edidiong Asikpo on September 18, at noon ET.

[![Getting Started with the DOM](/images/posts/DOM-getting-started.jpg)]
(https://cfe.dev/events/getting-started-with-the-dom/)
```

Figure 4.4 Prose provides a Markdown editor with simple preview capabilities and front matter metadata editing for any Markdown file in any repository within your GitHub account.

NETLIFY CMS

Netlify CMS (https://www.netlifycms.org/), shown in figure 4.5, is an open source content management tool. One of its differentiators is the ability to configure it to work with just about any static site generator. It was built by and is largely maintained by Netlify, so some features work out-of-the-box with Netlify's hosting, but it can be configured to work with other providers as well.

4.2.3 Why Netlify CMS?

Our project is going to be a technical documentation for the esoteric programming language LOLCODE. For this project, we're going to use Netlify CMS. There are a number of reasons this is a good choice:

- Netlify CMS is a liberally licensed open source project (it uses an MIT license). This means we can make our entire documentation project open source without worrying about license restrictions.
- Netlify CMS has an open authoring feature that allows us to give anyone with a GitHub account access to make contributions to the documentation. The users will have full access to the content management system, but their contributions will be automatically submitted as pull requests on their behalf, meaning the changes will not affect the site until we have accepted them.
- There is no limit to the number of users who are allowed to make a contribution and no costs associated with the number of users.

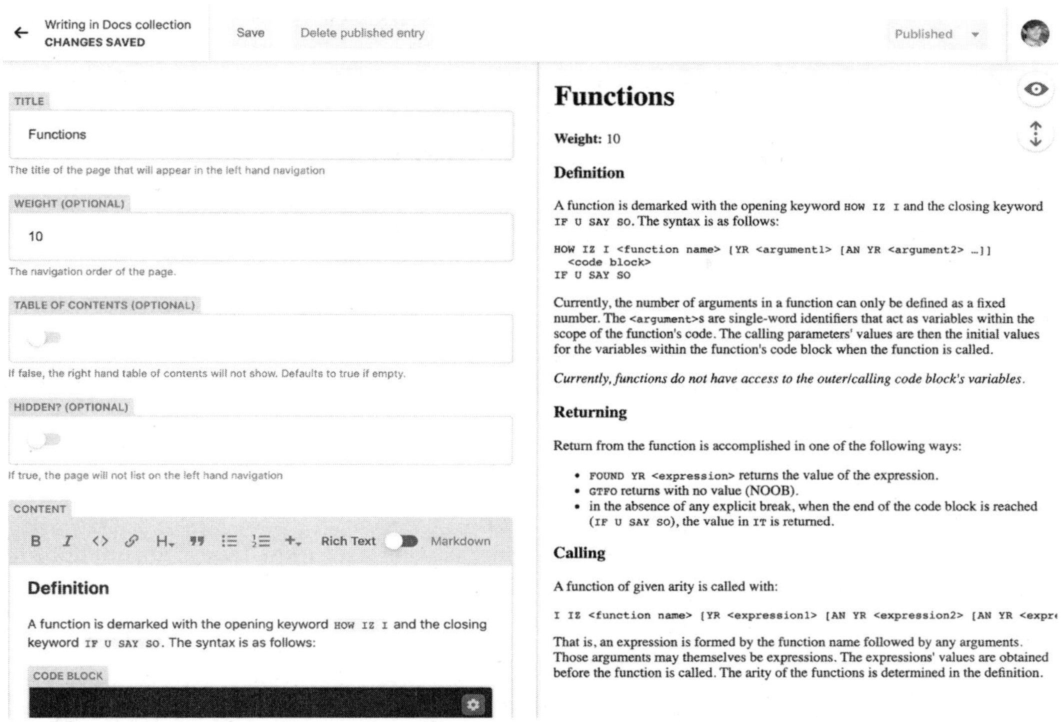

← Writing in Docs collection
CHANGES SAVED
Save Delete published entry Published

TITLE
Functions
The title of the page that will appear in the left hand navigation
WEIGHT (OPTIONAL)
10
The navigation order of the page.
TABLE OF CONTENTS (OPTIONAL)
If false, the right hand table of contents will not show. Defaults to true if empty.
HIDDEN? (OPTIONAL)
If true, the page will not list on the left hand navigation
CONTENT
B I <> H ”” Rich Text Markdown
Definition
A function is demarked with the opening keyword `HOW IZ I` and the closing keyword `IF U SAY SO`. The syntax is as follows:
CODE BLOCK

Functions

Weight: 10

Definition

A function is demarked with the opening keyword `HOW IZ I` and the closing keyword `IF U SAY SO`. The syntax is as follows:

```
HOW IZ I <function name> [YR <argument1> [AN YR <argument2> …]]
  <code block>
IF U SAY SO
```

Currently, the number of arguments in a function can only be defined as a fixed number. The `<argument>`s are single-word identifiers that act as variables within the scope of the function's code. The calling parameters' values are then the initial values for the variables within the function's code block when the function is called.

Currently, functions do not have access to the outer/calling code block's variables.

Returning

Return from the function is accomplished in one of the following ways:

- `FOUND YR <expression>` returns the value of the expression.
- `GTFO` returns with no value (NOOB).
- in the absence of any explicit break, when the end of the code block is reached (`IF U SAY SO`), the value in `IT` is returned.

Calling

A function of given arity is called with:

```
I IZ <function name> [YR <expression1> [AN YR <expression2> [AN YR <expr
```

That is, an expression is formed by the function name followed by any arguments. Those arguments may themselves be expressions. The expressions' values are obtained before the function is called. The arity of the functions is determined in the definition.

Figure 4.5 Netlify CMS has a variety of advanced widgets for editing both the long-form Markdown content of a post as well as the front matter metadata.

4.2.4 Static site generator (SSG) options

Any SSG can work for a documentation site. That being said, there are actually a number of SSGs that are specifically geared toward documentation:

- *Docsify* (https://docsify.js.org)—A Javascript-based SSG that serves documentation as a single-page application (SPA). Docsify differs from other solutions in that it does not generate static HTML files but instead parses the Markdown content at runtime in the browser, meaning the application does not need to rebuild to reflect changes or new content.
- *Slate* (https://github.com/slatedocs/slate)—A Ruby-based solution built on the Middleman SSG under the covers that is designed specifically for creating API docs. It also runs as a SPA that supports code examples in multiple languages, allowing the user to switch to the language tab that is relevant to them.
- *MkDocs* (https://www.mkdocs.org/)—A Python-based solution that emphasizes its speed when generating a large number of pages. It offers many themes, including a lot of community-built themes.

- *Docusaurus* (https://docusaurus.io/)—A JavaScript-based static site generator that uses React, Docusaurus comes with a lot of documentation-focused features and layouts out of the box, including things like documentation versioning and internationalization (i18n).
- *Hugo* (https://gohugo.io/)—A popular Go-based SSG that is also focused on an extremely fast build process, including built-in asset management. While it isn't documentation specific, Hugo has a large community of users with a lot of community-built themes, many of them specifically designed for documentation.

4.2.5 Why Hugo?

We're going to use Hugo for our LOLCODE technical documentation project. It will handle our growing volume of content easily while keeping build times down. Hugo is installed via a binary, meaning there's no complicated environment required for contributors who wish to run the project locally. It also has extensive and detailed documentation as well as a large number of community posts, making it easy to find solutions to any potential problems we may encounter.

Even though it isn't a documentation-specific solution, Hugo has a significant number of community themes that offer designs and features targeted to documentation sites. Some examples include the following:

- Ace Documentation is a Bootstrap-based docs theme.
- DocuAPI is geared toward multilingual API documentation.
- Dot is aimed at documentation in the form of a support center or knowledge base.
- Hugo Book is a minimalist book-style theme with features like built-in search.
- Techdoc is also a minimalist book-style theme.
- Kraiklyn is designed for creating single-page documentation.

For our example, we'll choose the Hugo Book theme. I chose this not for any technical reason but because I think the simple, clean layout it provides works well for the sort of language documentation we are creating.

4.3 Building the example site

Let's get started building our documentation site. We'll begin by installing Hugo and getting our theme set up and then configure the site to work with Netlify CMS.

4.3.1 Installing Hugo

There are a number of ways to install Hugo, including simply downloading the binary. While that is a method that works across all supported platforms (MacOS, Windows, and Linux), it has some complications in that you'll want to place it on your `path` so that it can easily be called from any location with just the `hugo` command-line command (rather than the full path to the binary). It's important to note that if you choose the binary install or already have Hugo installed, you need the extended version of Hugo version 0.68 or higher.

It is preferable to use a package manager to install Hugo.

Installing on MacOS or Linux

You can use Homebrew to install Hugo on MacOS or Linux:

```
brew install hugo
```

Installing on Windows

You can use Chocolatey to install Hugo on Windows:

```
choco install hugo-extended -confirm
```

Confirming your installation

Confirm that your Hugo installation worked properly:

```
hugo version
```

This should return something like the following (note that the version will have changed since this writing):

```
Hugo Static Site Generator v0.74.3/extended darwin/amd64 BuildDate: unknown
```

4.3.2 Creating a new Hugo site

We're going to build a technical documentation site for the LOLCODE esoteric language using Hugo as our static site generator, Netlify CMS as our Git-based headless CMS, and Hugo Book as our site's theme. The first step will be to generate a new site skeleton for our site using Hugo.

To create a new site using Hugo, use the `new site` command followed by a name for the directory you want the site to be created in:

```
hugo new site lolcode-docs
cd lolcode-docs
```

This will create a skeleton for a Hugo site with the following contents:

```
├── archtetypes
│   └── default.md
├── content
├── data
├── layouts
├── static
├── themes
├── config.toml
```

As you can see, the skeleton has no default content or theme. For the most part, Hugo generates only the directory structure and a basic configuration file. Here's what each of these files and folders are used for:

- In Hugo, archetypes represent the different content types in your application. These are templates for the front matter (i.e., metadata) of the different types of content your site will contain. For example, you may have a post archetype for blog posts that define the front matter a blog post will contain. While it isn't necessary that you create an archetype for all your content, doing so allows you to use the `hugo new` command with that type to generate a new page with the correct settings for that content type. So, if we had a post type, we could enter the command `hugo new post/my-new-post.md` to create a blog post with the name "my-new-post."
- All content for a Hugo site exists within the content folder. This can be in any directory structure within that folder. When the site is generated, a page will be created for every content item within the content folder. For example, a Markdown file at /content/posts/my-new-post.md will generate a page at /posts/ my-new-post/ within the site.
- The data folder contains all data files (YAML, TOML, or JSON). Hugo makes these available to the site within the `.Site.Data` object. For instance, a data file of authors.yaml would be available as `.Site.Data.Authors`.
- The layouts folder contains all the layout templates that Hugo will use to generate the pages. Typically, this folder is used on a site that does not have a theme installed. If both exist, Hugo will use the more specific file in layouts first (we'll use this to our advantage later). Hugo layouts are written using the Go template language to generate markup.
- The static folder contains any files that should be moved to the site without processing. These are often assets like images, JavaScript, or stylesheets that you do not want Hugo to modify. Everything from the static folder will be placed in the site root. For example, if you have a /static/images folder filled with the site's images, those will end up in just /images on the site. To show you what I mean, download the LOLCODE logo and save it in /static/images (the resulting file should be /static/images/logo.png).
- The themes folder is where you would place third-party themes that you download. You can find a ton of these at themes.gohugo.io. You can also create your own theme in this directory. To set the site's theme, you'll need to define a theme variable in Hugo's config.toml (we'll look at this in a moment).
- The config.toml is Hugo's configuration file, written in TOML. The base configuration Hugo provides includes just a `baseUrl`, `languageCode`, and `title`.
- Let's go ahead and drop in some default content for our docs site. I have provided a zip file containing the Markdown content for the sample site in the

book's GitHub repository at http://mng.bz/1jRy. Download that zip and extract it to the /content folder of your new site. You should now have /content/_index.md, which is the home page, and /content/docs, which will contain a number of Markdown files with the site documentation.

4.3.3 Setting up the Hugo Book theme

We're going to install the Hugo Book theme as a submodule. Before we can do that, we'll need to ensure our new project is initialized as a Git repository. Using the terminal/command line at the root of your project directory, initialize a new repository:

```
git init .
```

Next, let's add hugo-book as a submodule. Installing the theme this way will allow us to keep our project up to date if any changes are made to the theme in the GitHub repository:

```
git submodule add https://github.com/alex-shpak/hugo-book themes/book
```

Finally, let's configure Hugo to use the newly installed theme. We're going to open the `config.toml` in the root folder of the project and make the following changes:

- Change the title to "LOLCODE Documentation."
- Add a theme variable to set the theme to "book."
- Add additional Hugo Book configuration. The theme provides an extensive amount of configuration options, but we'll just add the ability to search using the `BookSearch` parameter.

Here's what the finished configuration file looks like:

```
baseURL = "http://example.org/"
languageCode = "en-us"
title = "LOLCODE Documentation"
theme = "book"
[params]
    BookSearch = true
```

For now, we're leaving the `baseURL` value alone. This value represents the hostname and path to the root of the site and can be used in Hugo layout code. Once we've deployed our site to Netlify, we can update this, but for now it won't impact our project.

Now we're ready to test our site. Just as a reminder, be sure you downloaded the LOLCODE logo and saved it in /static/images. From the terminal/command line, run `hugo serve` from within the project's root folder to launch Hugo's local web server. This will build your site and make the page viewable at http://localhost:1313. If you open the site in your browser, it should look like figure 4.6.

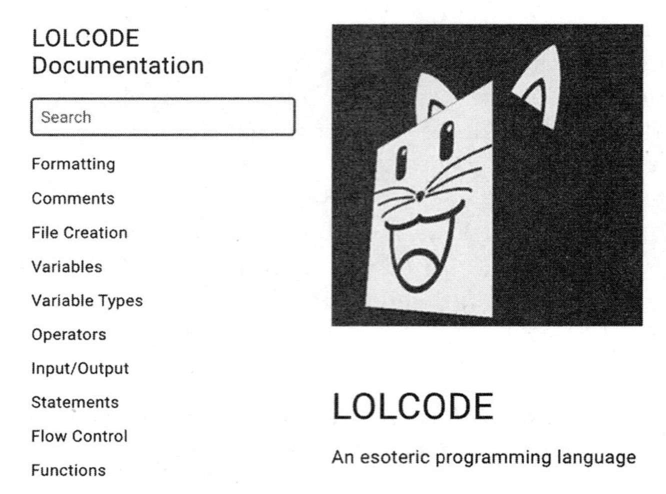

Figure 4.6 The LOLCODE documentation site running on our local Hugo web server

Feel free to browse around or even search. You'll see that we now have a fully functional documentation site. We could choose to leave it as-is and simply maintain content in GitHub, but, as we noted earlier, we want to include a CMS and allow third-party contributors access. Let's do that.

4.3.4 *Installing Netlify CMS*

Now that we have content and a functioning site, we can enable it to be edited with Netlify CMS. Before we do that, we need to ensure we've published our project on GitHub. Please do this if you have not already.

There is no installer for Netlify CMS. Instead, you create the admin and add the necessary files into it. Here are the steps:

1. Create a folder named `admin`. Since this is Hugo, we need to place it within the static directory (i.e., /static/admin) because we do not want the files to be processed by Hugo.
2. Add an index.html within /static/admin that will load the script that runs the CMS admin. We'll use the code supplied by the Netlify CMS docs:

```
<!doctype html>
<html>
<head>
  <meta charset="utf-8" />
  <meta name="viewport" content="width=device-width,
  ➥ initial-scale=1.0" />
  <title>Content Manager</title>
</head>
<body>
```

```
    <!—Include the script that builds the page and powers Netlify CMS ->
    <script s"c="https://unpkg.com/netlify-cms@^2.0.0/dist/netlify-
    ➥ cms"js"></script>
  </body>
  </html>
```

3 Create a config.yml within /static/admin that will contain the Netlify CMS configuration. This will eventually contain our full content model definition, but for now we'll add only the base configuration information that we need (note that you will need to replace remotesynth/lolcode-docs with your GitHub repository information):

```
publish_mode: editorial_workflow
media_folder: static/images
public_folder: /images
backend:
  name: github
  branch: master
  repo: remotesynth/lolcode-docs
  open_authoring: true
```

Let's review what's in this configuration file:

- Setting `publish_mode` to `editorial_workflow` creates a draft, review, and approve workflow for content. Without setting this, content would be automatically published on save. The editorial workflow is a requirement of enabling open authoring.
- The media_folder defines the folder within the site source where images and other media files can be uploaded. The public_folder then defines the path to the media_folder within the published site.
- The backend section defines the repository details that Netlify CMS uses as the site's backend. The default is Git Gateway, which is an open source API that proxies requests between users on your site and the Git repository. This works with Netlify Identity, Netlify's authentication solution, out of the box. However, for open authoring in Netlify CMS we must use GitHub, which uses GitHub's OAuth authentication to allow users access. We then define the branch and the repository for the site (the sample code points to my repository, so make sure the repository reflects your GitHub repository). Finally, we set `open_authoring` to true to allow external contributors without needing to invite them.
- We've configured the basic settings for Netlify CMS, but it's not going to work just yet for two reasons: we have not modeled any content for Netlify CMS, so it does not know anything about what it is editing, and we have not set up either Netlify or GitHub to allow users to authenticate.

4.3.5 Modeling content in Netlify CMS

Before it can begin editing your content, Netlify CMS needs to understand its structure. It does this by defining `collections` and `fields` within the config.yml. Depending on the complexity of your site, modeling content for Netlify CMS can become a pretty involved task. Luckily, our documentation site's content is pretty straightforward.

COLLECTIONS

A collection is a content type in Netlify CMS. This can represent a single page, a group of pages with common attributes, or a data file (i.e., YAML, TOML, or JSON). There are two collection types in Netlify CMS:

- A *folder collection* represents a group of content files that all reside within a single folder. It is important to note that, as of this writing, Netlify CMS does not support subfolders, meaning that if you had /content/docs/topic-one and /content/docs/topic-two, they could not all be defined using the /docs folder and would require three separate collection definitions.
- A *files collection* represents one or more single files: a page (or pages) in Markdown or HTML or a data file (or files) in YAML, TOML, or JSON. When referencing a page in Markdown or HTML, you'd typically use the file type for a specialized page that does not share attributes with any other of the site's pages, for example, the site's home page.

Our documentation site has two content types, one representing the home page, which is a file type, and one representing the docs, which is a folder type. Place this beneath the `backend` configuration block right after the line containing `open_authoring: true`:

```
collections:
  - name: pages
    label: Pages
    files:
      - nam“: "h”me"
        labe“: "Home P”ge"
        fil“: "content/_index”md"
  - name: docs
    label: Docs
    folder: /content/docs
    create: true
    extension: md
    slug: '{{slug}}'
```

Let's explore what we've configured so far:

- Every collection must be given a `name` that is a unique identifier of the collection within Netlify CMS. You can use any name you choose, but you should avoid spaces or special characters other than dashes or underscores. Meanwhile, the `label` defines how the collection will be displayed to the user within the CMS. You can label it however you like.

- A folder collection represents a single folder with multiple files. Users will have the ability to create new pages (`create` is set to `true`), and these new pages will be Markdown with a file extension of `md`. The `slug` field defines how Netlify CMS will generate new file names. In our case, we are saying to generate a URL-safe version of the content's title (this means our field definition must contain a `title` field).
- A files collection must define the different specific files it contains. There can be multiple, and each can define their own fields (we'll discuss this shortly). This means that each file in a files collection does not need to share properties, but the collection is a way of grouping them together from an editing perspective. Our documentation site has only one file representing the home page.

Our configuration touches on only a small fraction of the options available to you. Check the documentation for a full list of collection configuration options (http://mng.bz/7W9m).

FIELDS

Fields represent the different data properties (metadata) on a content object. For example, a blog post might have a title and a date property, among others, that need to be defined as fields within Netlify CMS.

Each field is represented by a *widget*. A widget in Netlify CMS determines how this particular field will be edited. For example, a `text` widget would be an HTML `textarea` field, a `boolean` widget would be a toggle switch, and an `image` widget would be a file picker. Netlify CMS comes with 16 default widgets that cover most use cases, but you can define your own custom widgets as well.

Each field we define in our content model has the following common properties:

- A `widget` property that defines which widget will be used for this field in the CMS user interface.
- A `name` that is the field name within Netlify CMS and should be unique within this group of fields. You can name it anything, but avoid spaces or special characters other than dashes or underscores.
- A `required` attribute to specify if the field is required. If this isn't included, the default is true.
- A `hint` field that defines text that will appear in a tool tip when the widget is displayed in the CMS user interface. This is optional and can be used to offer additional help or context to the user for entering values.
- A `pattern` field that can define a regular expression (regex) pattern for validating the input and an error message to display when the validation fails. This is optional.

In addition, each type of widget can have widget-specific configuration properties. Check the documentation for the full list of options (https://www.netlifycms.org/docs/widgets/).

Our field definitions are all fairly straightforward, as the content model for our documentation isn't particularly complex. Here's the full complete configuration file with collections and fields (be sure to update the `repo` with your own GitHub repository).

Listing 4.1 The completed Netlify CMS configuration file (/static/admin/config.yml)

```
publish_mode: editorial_workflow
media_folder: static/images
public_folder: /images
backend:
  name: github
  branch: master
  repo: remotesynth/lolcode-docs
  open_authoring: true
collections:
  - name: pages
    label: Pages
    files:
      - name: "home"
        label: "Home Page"
        file: "content/_index.md"
        fields:
            - widget: string
              name: title
              label: Title
              required: true
              hint: >-
                The title of the page
            - widget: markdown
              name: body
              label: Content
              required: true
              hint: Page content
  - name: docs
    label: Docs
    folder: /content/docs
    create: true
    extension: md
    slug: '{{slug}}'
    fields:
        - widget: string
          name: title
          label: Title
          required: true
          hint: >-
            The title of the page that will appear in the left hand
            ➥ navigation
        - widget: number
          name: weight
          label: Weight
          required: false
```

```
      hint: >-
        The navigation order of the page.
    - widget: boolean
      name: bookToc
      label: Table of Contents
      required: false
      hint: >-
        If false, the right hand table of contents will not show.
        ➥ Defaults to true if empty.
    - widget: boolean
      name: bookHidden
      label: Hidden?
      required: false
      hint: >-
        If true, the page will not list on the left hand navigation
    - widget: markdown
      name: body
      label: Content
      required: true
      hint: Page content
```

With the configuration in place, we should be able to run `hugo serve` from the command line and then navigate to http://localhost:1313/admin and see a login for the admin interface.

Clicking Login with GitHub will not yet work (figure 4.7), as we have not configured Netlify or GitHub for authentication. Let's do that next.

Figure 4.7 The Netlify CMS login at /admin using GitHub for user authentication

4.3.6 Deploying to Netlify

It is possible to use Netlify CMS without deploying to Netlify, but since Netlify created the project, it has the most straightforward path when it comes to enabling the authentication that will allow users access to editing the content via the CMS. We'll look at deployment in depth in a later chapter, but for now we'll cover the basics needed to allow users in our Netlify CMS admin.

First, make sure you've published your project to GitHub. Again, Netlify CMS allows for other Git hosting providers, but we'll be using GitHub for authentication, so publishing there is a requirement in this scenario. Since the repository will be editable by third parties, be sure to make the repo public.

You'll need to create an account on Netlify if you don't already have one. Netlify offers a generous free plan that will enable you to complete this tutorial. Once you've created your Netlify account, click the New Site from Git, choose GitHub, and then locate your published repository. If this is your first time using Netlify, you will need to walk through some authorization steps to allow Netlify access to your GitHub repositories, as shown in figure 4.8.

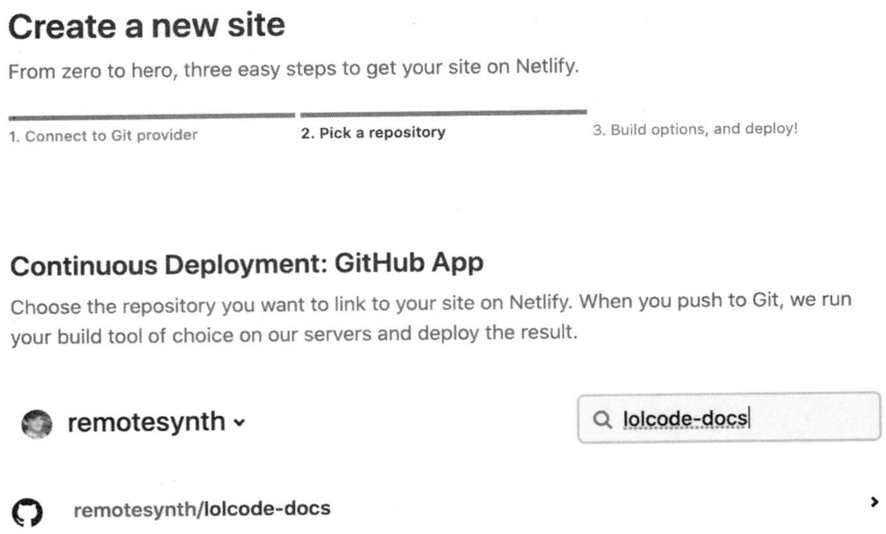

Create a new site

From zero to hero, three easy steps to get your site on Netlify.

1. Connect to Git provider 2. Pick a repository 3. Build options, and deploy!

Continuous Deployment: GitHub App

Choose the repository you want to link to your site on Netlify. When you push to Git, we run your build tool of choice on our servers and deploy the result.

remotesynth lolcode-docs

remotesynth/lolcode-docs

Figure 4.8 Creating a new site from a GitHub repository in Netlify

Netlify does a good job of picking up on which static site generator we're using and the default settings for it. However, I have frequently run into issues when using a recent build of Hugo, so I find it best to set an environment variable that matches the Hugo version you are running locally. From the command line, type `hugo version` to see what version you are running. For example, mine returns the following:

```
Hugo Static Site Generator v0.74.3/extended darwin/amd64 BuildDate: unknown
```

Click on the Show advanced button in the deploy settings step of the setup, as seen in figure 4.9.

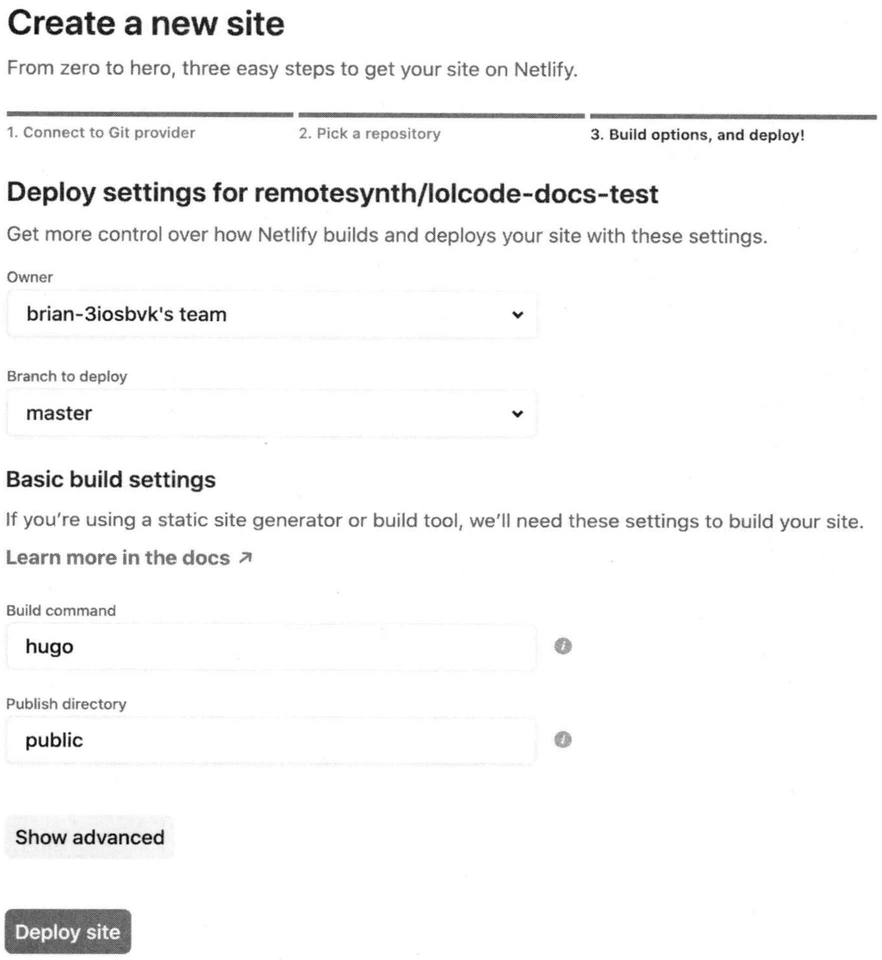

Create a new site

From zero to hero, three easy steps to get your site on Netlify.

1. Connect to Git provider
2. Pick a repository
3. Build options, and deploy!

Deploy settings for remotesynth/lolcode-docs-test

Get more control over how Netlify builds and deploys your site with these settings.

Owner

brian-3iosbvk's team

Branch to deploy

master

Basic build settings

If you're using a static site generator or build tool, we'll need these settings to build your site.

Learn more in the docs ↗

Build command

hugo

Publish directory

public

Show advanced

Deploy site

Figure 4.9 The default deploy settings when creating a Hugo site in Netlify

Click the New Variable button and then add a variable named `HUGO_VERSION` with a value of the version number that is returned by running `hugo version`. For example, in my case, the version is 0.74.3, which I've entered in figure 4.10.

Finally, click on Deploy Site. After a few minutes, your site should be deployed. Grab the URL that Netlify generated for your site (this can be found in the Site overview page in the Netlify dashboard). Let's fix the `baseURL` value in the config.toml file in the root of the project by setting its value to the URL on Netlify. For example, mine is https://clever-thompson-493f7c.netlify.app/. This will fix any missing stylesheets you may see on the site when it is initially deployed. We'll also need the URL for configuring authentication in GitHub.

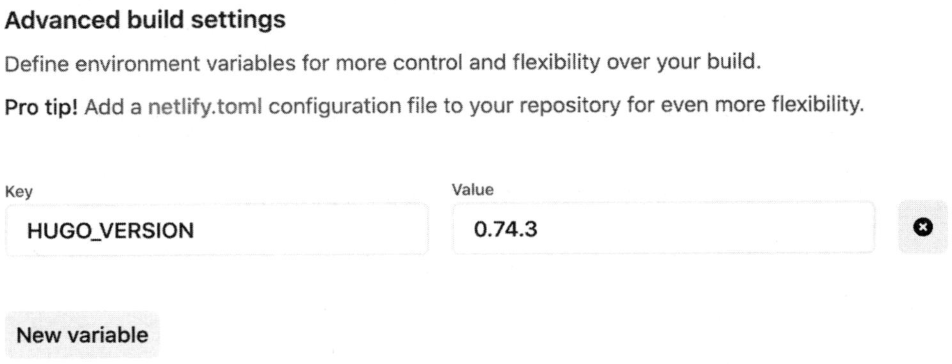

Advanced build settings

Define environment variables for more control and flexibility over your build.

Pro tip! Add a netlify.toml configuration file to your repository for even more flexibility.

Key: HUGO_VERSION
Value: 0.74.3

New variable

Figure 4.10 Setting the Hugo version variable in Netlify's deploy settings setup

4.3.7 *Configuring GitHub for authentication*

Before we can set up authentication in Netlify, we'll need to set up an OAuth application in GitHub. To do this, go to Settings > Developer Settings > OAuth Apps and click the button that reads either Register a new application or New OAuth App or visit https://github.com/settings/applications/new.

We need to give our new OAuth App a name; it can be anything you want. In the Homepage URL field, place the URL of your Netlify site (which we received in the prior section). The description can also be anything you want. Finally, the authorization callback URL needs to be https://api.netlify.com/auth/done. You can see these settings in figure 4.11.

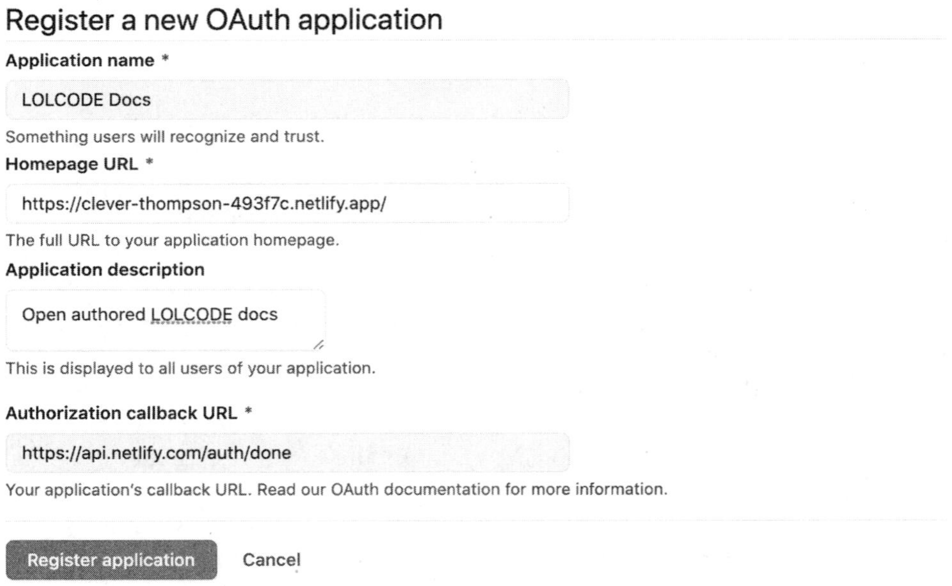

Register a new OAuth application

Application name *
LOLCODE Docs
Something users will recognize and trust.

Homepage URL *
https://clever-thompson-493f7c.netlify.app/
The full URL to your application homepage.

Application description
Open authored LOLCODE docs
This is displayed to all users of your application.

Authorization callback URL *
https://api.netlify.com/auth/done
Your application's callback URL. Read our OAuth documentation for more information.

Register application   Cancel

Figure 4.11 Setting up a new OAuth app in GitHub that can be used with Netlify's authentication

After you click Register application, we're given a client ID and client secret for our OAuth application. We will need these to set up Netlify.

4.3.8 Configuring Netlify for authentication

In our Netlify dashboard for our new site, we need to go to Site Settings > Access Control > OAuth. Click the Install Provider button.

As seen in figure 4.12, the provider should be GitHub and, in the client ID and secret fields, place the client ID and client secret we received from our GitHub OAuth application. Lastly, click Install.

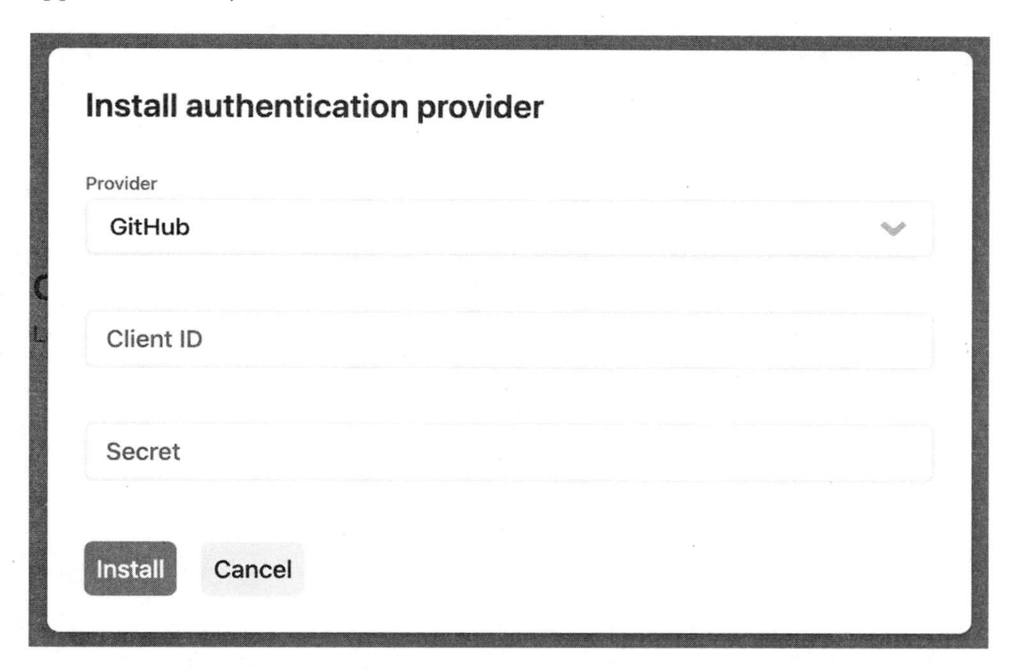

Figure 4.12 Adding an OAuth provider in Netlify. The client ID and secret come from the OAuth application we created in GitHub.

4.3.9 Editing content as an admin

We're now ready to access the content admin on our documentation site (be sure you've pushed any changes we've made to GitHub first). The admin is available at /admin. For instance, the URL of my Netlify site is https://clever-thompson-493f7c.netlify.app/, so the URL for my Netlify CMS admin will be https://clever-thompson-493f7c.netlify.app/admin. Before moving on, be sure that you have your GitHub repository set for the value of `repo` in the Netlify CMS configuration (/admin/config.yml).

Click the Login with GitHub button, and we receive the authorization window for GitHub based on the GitHub OAuth app we created, as seen in figure 4.13.

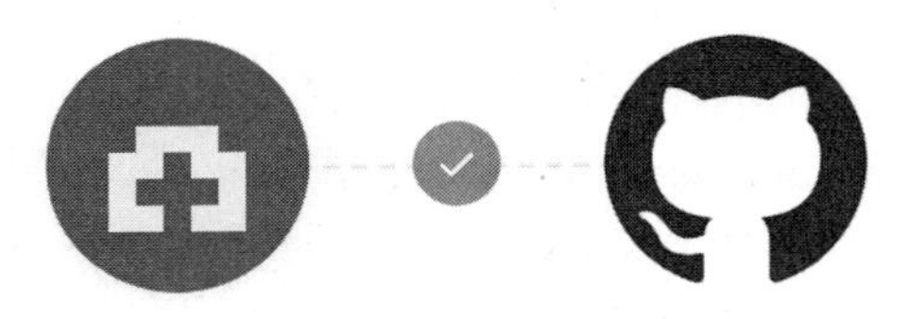

Figure 4.13 The GitHub authorization window is displayed the first time you click Login with GitHub and displays the information we entered when creating our GitHub OAuth app.

Once we click Authorize, we're logged in and taken to the Netlify CMS editing dashboard, which you can see in figure 4.14.

By default, we're in the Contents tab that shows the content collections we defined earlier in the config.yml configuration file: Pages and Docs. The Pages collection is selected by default. You may recall that this collection had only one content item defined, which is the home page content. You cannot add new pages into the Pages collection.

Figure 4.14 The Netlify CMS dashboard after logging in. The collections are the ones we defined in the Netlify CMS configuration.

When we click the Docs collection, we see a full list of documentation pages as well as a button to create a new docs page. Feel free to click one to edit it. The editing page is shown in figure 4.15.

Writing in Docs collection
CHANGES SAVED
Save
Delete published entry
Published

TITLE
Comments
The title of the page that will appear in the left hand navigation

WEIGHT (OPTIONAL)
2
The navigation order of the page.

TABLE OF CONTENTS (OPTIONAL)
If false, the right hand table of contents will not show. Defaults to true if empty.

HIDDEN? (OPTIONAL)
If true, the page will not list on the left hand navigation

CONTENT
Rich Text Markdown

(from 1.1)
Single line comments are begun by BTW, and may occur either after a line of code, on a separate line, or following a line of code following a line separator (,).
All of these are valid single line comments:
CODE BLOCK

Comments

Weight: 2

(from 1.1)

Single line comments are begun by BTW, and may occur either after a line of code, on a separate line, or following a line of code following a line separator (,).

All of these are valid single line comments:

```
I HAS A VAR ITZ 12          BTW VAR = 12
I HAS A VAR ITZ 12,         BTW VAR = 12
I HAS A VAR ITZ 12
              BTW VAR = 12
```

Multi-line comments are begun by OBTW and ended with TLDR, and should be started on their own lines, or following a line of code after a line separator.

These are valid multi-line comments:

```
I HAS A VAR ITZ 12
            OBTW this is a long comment block
                 see, i have more comments here
                 and here
            TLDR
I HAS A FISH ITZ BOB

I HAS A VAR ITZ 12, OBTW this is a long comment block
      see, i have more comments here
      and here
TLDR, I HAS A FISH ITZ BOB
```

Figure 4.15 Editing one of the Docs pages in Netlify CMS. We can see the widgets for all the content properties we defined in the fields in the config.yaml.

Each of the widgets on the left-hand side of the page represents the properties we defined earlier for the `docs` content type in the config.yml configuration file. The content field offers a WYSIWYG-style editing interface for Markdown content. The right side of the page offers a preview of the content being edited.

Go ahead and make some changes to the content. Once you've made changes, click the Save button. Since we are using an editorial workflow, the document will be saved as a draft in the workflow. We can change the status to In Review or Ready. We'll need to set the status to Ready before we can publish our changes to the page.

In addition, by using Netlify's deploy previews feature, we can preview our changes on the site before we publish. In the top bar of the page, we'll see a "Check for preview" link. Clicking that link will take us to the link for the deploy preview of our changes from Netlify, which may take a few seconds. Clicking View Preview will open the deploy preview of the site and will include our changes so that we can review them before publishing.

When you are ready, change the status to Ready, click Publish, and choose Publish Now. This will commit the changes to our GitHub repository, which will then trigger a build of our Netlify site and publish the changes to our live site.

4.3.10 The open authoring workflow

The flow for our external users will be slightly different. Let's explore what this looks like. You don't need to follow along with this part, as it would require a secondary GitHub account.

Once they log in with GitHub and authorize our GitHub OAuth app, they will be asked to fork the repository as seen in figure 4.16. Clicking Don't fork the repo at this point will exit the process, and the user will be unable to make any edits.

Clicking Fork the repo will automatically create a fork of our repository on the user's account. This is where all of the changes the user makes will be kept.

Figure 4.16 External users will be asked to fork the repo in order to get access to the Netlify CMS admin and submit edits to the site.

Once the user forks the repo, the Netlify CMS admin will be identical to the one we used to edit our site earlier. However, when they make changes to the site, they do not have the option to set the status to Ready or any of the Publish options. Instead, they will only have the option to set the status to In Review. Doing so will automatically submit a pull request to our main repo that contains the changes the user made, as you can see in figure 4.17.

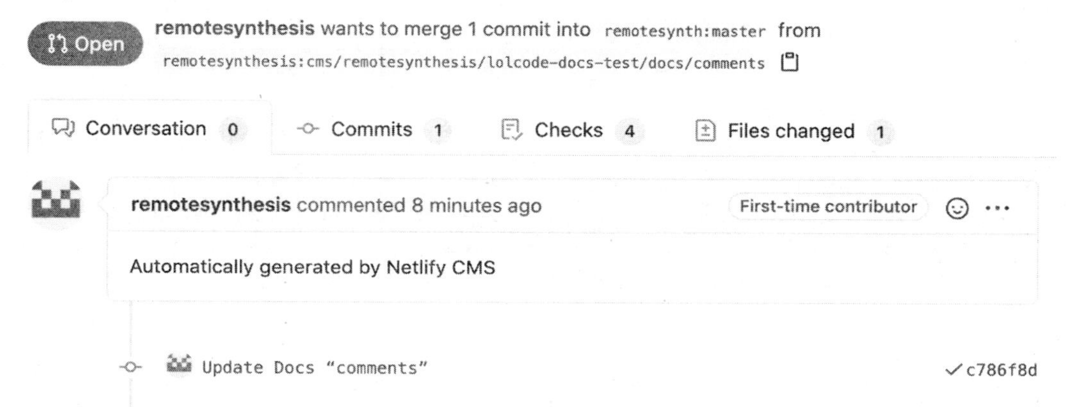

Figure 4.17 When a third-party contributor sets a change as In Review, a pull request is automatically submitted on the primary GitHub repository. To accept the change, we can merge the pull request.

To accept the user's change into our site, we need to merge the pull request. This updates our GitHub repository, triggers an update on our Netlify site, and publishes the changes to our live site.

4.3.11 Simplifying the open authoring workflow

Before we finish, let's make one last edit to our site. Right now, any external author wishing to contribute to the docs needs to know to go to /admin and log in. Contributions would be much more likely if we were to give users quick links for adding or editing content from within the documentation itself. To do this, we'll make a small change to the template.

We are relying on the Hugo-Book theme for our layout and have installed this as a Git submodule. Thus, we cannot change the Hugo-Book code directly. Nonetheless, Hugo has a lookup order for theme templates that takes the most specific match first. This means that we can place files in the /layouts folder that can override the template files in the /themes folder. This will make more sense when we see how this works in practice.

The Hugo-Book theme has layout files that are specifically intended to inject content at certain points in the template output, such as before or after the content. You can see these files in /themes/book/layouts/partials/docs/inject. Let's add the links at the end of each content page via the `content-after.html` template.

To override this template, create a file in your /layouts folder with the same directory structure and name: /layouts/partials/docs/inject/content-after.html. Since Hugo considers a file in /layouts more specific than the one in /themes, by creating a file of the exact same path, Hugo will use it in place of the theme file.

Place the following template code inside the file:

```
{{ if ne .RelPermalink "/"}}
    {{ $edit_url := print "/admin/#/edit" .RelPermalink }}

    <p><a href="{{$edit_url}}" class="book-btn">Edit this
     Page</a>  
    <a href="/admin/#/collections/docs/new" class="book-btn">Add a New
    ➥ Page</a></p>
{{ end }}
```

Let's look at what this template code is doing. First, it checks to see if we are on the home page by checking a variable Hugo provides called `.RelPermalink` that contains the relative path of the current page. The function `ne` in Hugo means "not equal," so we are checking that the relative path is not equal to "/." We do this so as not to offer edit links on the home page. Second, we assemble a URL to the admin based on the current page. Finally, we use that URL to add an Edit this Page link, which will take them directly into editing the content for the page they are currently viewing. We also include an Add a New Page link that directly links to creating a new docs page.

After committing and pushing this page to our repository or running it locally, we should see these new links added to any Docs page on our site, such as the one shown in figure 4.18.

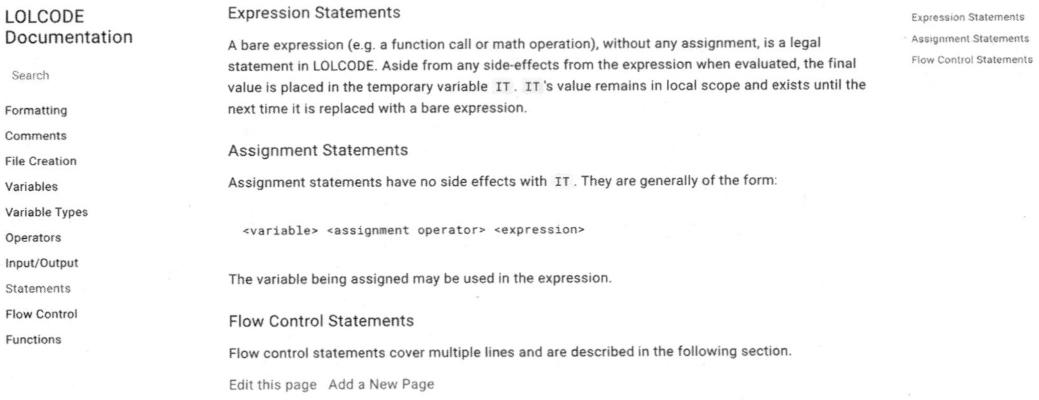

LOLCODE Documentation

Search

Formatting
Comments
File Creation
Variables
Variable Types
Operators
Input/Output
Statements
Flow Control
Functions

Expression Statements

A bare expression (e.g. a function call or math operation), without any assignment, is a legal statement in LOLCODE. Aside from any side-effects from the expression when evaluated, the final value is placed in the temporary variable `IT`. `IT`'s value remains in local scope and exists until the next time it is replaced with a bare expression.

Assignment Statements

Assignment statements have no side effects with `IT`. They are generally of the form:

```
<variable> <assignment operator> <expression>
```

The variable being assigned may be used in the expression.

Flow Control Statements

Flow control statements cover multiple lines and are described in the following section.

Edit this page   Add a New Page

Expression Statements
Assignment Statements
Flow Control Statements

Figure 4.18 Once we've added our template code, Docs pages will have Edit this Page and Add a New Page links.

4.4 What's next?

As we've seen, the Jamstack can be a powerful solution for documentation sites. An open source solution like Netlify CMS offers the ability to use a Git-based workflow while still allowing content editors an easy-to-use WYSIWYG editing experience, and even allowing third-party contributions—something that isn't easy with a non-Jamstack solution.

While the editor experience of Netlify CMS is full-featured, some may feel that it lacks the polish of some non-Jamstack tools like WordPress. It's worth keeping in mind that there are many alternatives that offer a different editing user experience. If you're looking for a more WordPress-like experience, be sure to explore the API-based headless CMS options like Contentful, Sanity, or AgilityCMS, or even services like Stackbit.

As we have shown, the Jamstack is an excellent solution for content-focused sites like documentation, but you may be wondering if it can handle a site with more complex and dynamic user interaction. In the next chapter, we'll look at just such an example as we build an e-commerce site using Jamstack tools.

Summary

- Documentation sites have been and continue to be a perfect use case for the Jamstack, as they are heavily focused on content and can benefit from things like versioning that are a core part of a Jamstack workflow.
- A headless CMS is a content management system that offers content editing tools that are independent of the site's frontend display. An API-based headless CMS makes the content available to the frontend via an API, while a Git-based headless CMS edits content directly in the site's Git repository.
- Netlify CMS is an open source, Git-based headless CMS created and maintained by Netlify, which offers the option of an open-authoring workflow. This can allow external contributors the option to edit and submit changes to the site's content.
- There are a lot of documentation-specific static site generators. While Hugo, a Go-based static site generator, is not documentation-specific, it is often a favored tool for these types of projects due to its build speed and a large number of available templates for documentation.
- Netlify CMS is configured via YAML and must have a model of the content on the site that the CMS will be able to edit. We configured Netlify CMS for the basic documentation content model used by our LOLCODE technical documentation site.
- Open authoring on Netlify CMS allows third-party contributors to have access to the CMS to make content contributions. We configured Netlify and GitHub for authentication to allow third parties to sign in to our CMS using open authoring.

- Netlify's admin interface uses widgets to enable easy editing of the page's metadata (front matter) and content. Markdown editing uses a WYSIWYG-style editing interface. The CMS user experience of both the site owner and third-party authors is nearly identical, other than the ability to mark updates as Ready and publish them, which is reserved for the site owner.

5 Building an e-commerce site

This chapter covers

- Establishing the needs of a typical e-commerce site
- Comparing headless e-commerce systems for managing products and checkout
- Choosing a static site generator for an e-commerce site
- Creating and configuring a new site using Next.js
- Building a product listing, product detail, and shopping cart in Next.js
- Importing and using Markdown content in a Next.js site

An e-commerce site has many requirements that might make it seem ill-suited for an architecture based on static assets, as the Jamstack is. While the content aspects, such as product listings and detail pages, can be made to fit easily within a statically generated site, things like shopping carts, checkout processing, and order histories appear too dynamic and interactive to function without server-side rendering.

Up until only a three or four years ago, this is exactly the advice I would give: static site generators were not a good fit for e-commerce sites. However, the Jamstack does more than just build sites with static site generators. A core piece of the Jamstack is its ability to use JavaScript and APIs on the client side to enable dynamic functionality that would otherwise be impossible using static site tooling. Thus, it is entirely possible today to build fully functioning e-commerce sites using the Jamstack.

Saying something is possible does not necessarily mean it is right. However, I'd argue that, in many if not most cases, adopting the Jamstack for e-commerce is the right idea. This is because there is a whole cottage industry of research showing the impact performance can have on conversion rates and, therefore, sales. For instance, one analysis found that "if a site makes $100,000/day, [a] one second improvement in page speed brings $7,000 daily" (http://mng.bz/Nxj7). In other words, the performance improvements that Jamstack offers will not only improve your users' experiences, but that improved experience can improve your bottom line.

5.1 Requirements of an e-commerce site

E-commerce sites can vary greatly in complexity. Some sites provide only a handful of products or services and a simple checkout process, without any real feature frills. Others offer all kinds of dynamic content, such as user reviews, personalized recommendations, wish lists, and more. All these features are possible using the Jamstack, and some of the tools that we'll discuss even offer special APIs for achieving them.

Let's look at a few core requirements for a typical e-commerce site:

- An e-commerce site has a list of products or services. This content should be easily updated and managed through some form of content management system to allow for quick and easy updates by nontechnical users tasked with maintaining the site.
- E-commerce sites have regular content pages beyond the product listings for things like the About page or terms of service. In most cases, these are infrequently updated, so they may not require integration within external content management.
- E-commerce sites generally have a shopping cart where users can dynamically add and remove items and proceed to the checkout process.
- An e-commerce site must have a checkout process to complete the purchase. In most cases this is integrated within the site, but in some cases the final checkout process can be offloaded to a third party for processing and confirmation.

5.1.1 The example site requirements

The example we're going to build for this chapter is a simple storefront, called the Jam Store, for selling toy figures and rubber ducks. It includes all the basic requirements discussed. Currently, we have a limited inventory of only four items, but we expect it to expand (see figure 5.1).

Figure 5.1 The final result of our example e-commerce project is a store that sells toy figures.

Our site will integrate with content management for adding and editing our product listings. However, our other content page—our About page—will be infrequently updated, so it will be managed as simply a Markdown file.

Users will be able to add items to their shopping cart from the product detail page. They'll be able to modify quantities or delete items from their cart via the My Cart page before proceeding to checkout. Since we're small (and for the sake of simplicity), the checkout process will be offloaded to a third party to manage final shipping and order confirmation details.

5.2 Choosing the right tools

There are myriad ways to build a Jamstack e-commerce site, including building a custom solution from scratch. However, there are a lot of complexities in e-commerce that can make this a bit of a daunting task. The Jamstack generally favors leveraging existing services wherever possible to simplify development, and in this case, we'll use a type of service called headless e-commerce.

5.2.1 What is headless e-commerce?

In chapter 4, we learned about a concept called a headless CMS, wherein the CMS provides the backend content management tools that are decoupled from the frontend presentation. Headless e-commerce is the same concept applied to e-commerce tooling. A *headless e-commerce solution* provides the tools for managing a shopping cart, orders, shipping, and, in many cases, products/services and inventory that are not tied to a specific frontend solution, allowing you to customize the frontend to suit your needs.

I'd argue there is even more variation in the types of headless e-commerce than in headless CMS, where we had just two primary types (API-based and Git-based). There are no commonly accepted categories of headless CMSs, but—borrowing from the work of François Lanthier Nadeau (https://snipcart.com/blog/headless-ecommerce)—CEO of SnipCart (one of the solutions we'll discuss), here are the broad types:

- *All-in-one solutions*—As the name implies, these provide full packages to manage every aspect of an e-commerce site, from content to products, to orders, shipping, and so on. In most cases, these tools are more geared toward a full-stack solution (coupled) that provides both the front- and backend for the site, but they also offer APIs to access the backend as a decoupled headless alternative that can be used in a Jamstack site.
- *Add-on solutions*—These tools offer a complete shopping cart and checkout solution that effectively lie on top of your site. This is typically done by including a script that embeds an overlay. These tools usually don't manage content and don't necessarily require that you manage your product listings through their service.
- *API-based solutions*—Like their headless CMS counterparts, all of the functionality from these services is available only via API calls. While the backend manages everything from products to shopping carts, orders, and shipping, they make no assumptions on how the frontend is built. Everything from adding and removing items from a cart to the checkout process is handled via calls to the API, using JavaScript in the case of a Jamstack site.

From a Jamstack perspective, both the all-in-one solutions and the API-based solutions will be consumed the same way: a Jamstack site will use the APIs provided by the all-in-one solution rather than rely on any of the frontend development tools.

The add-on solutions generally require the least amount of development effort, but the tradeoff is usually in ease of customizability. Once you've built a site with product/service listings, you can simply connect it with the headless e-commerce tool that manages the rest. On the other hand, the all-in-one and API-based solutions both require more development effort, as the frontend for things like the cart and checkout need to be custom built, but by relying on only APIs, the developer can create the frontend however they choose.

TIP For a more in-depth look at a range of headless e-commerce options, check out this detailed post by Bejamas: https://bejamas.io/blog/jamstack-ecommerce/.

5.2.2 Headless e-commerce options

Before we get into which option we'll choose, let's look at a few of the most popular headless e-commerce solutions, one from each category.

SHOPIFY

Shopify (https://shopify.dev/) is one of the most popular all-in-one e-commerce options on the market. Shopify's services can be used in a Jamstack site via their Storefront API (https://shopify.dev/api/storefront). This is a GraphQL API that offers access to the full range of Shopify's services for product information, orders, and checkout. Shopify also offers a JavaScript SDK (http://mng.bz/Dxja) for the Storefront API, which simplifies the code needed to interact with their API. Butcher Box (see figure 5.2) and Victoria Beckham Beauty are two storefronts built with the Jamstack and that access the Shopify's Storefront API.

Figure 5.2 Butcher Box is an example of an online store built with the Jamstack and that uses Gatsby to access the Storefront API.

SNIPCART

Snipcart (https://snipcart.com/) is an example of an add-on e-commerce solution. At its most basic, enabling a site to work with Snipcart only requires that it include the Snipcart JavaScript and CSS files and then add custom HTML attributes to an Add to Cart button. That's it. The link will trigger the Snipcart cart overlay to appear with all of the cart management and checkout features built in. Snipcart does include a number of customization options as well as a JavaScript SDK if you want to access any of its features programmatically rather than via the embedded cart and checkout.

> **NOTE** For a quick tutorial on setting up a Jamstack site built with Hugo that uses Snipcart for e-commerce functionality, check out this tutorial and sample app: https://www.stackbit.com/blog/ecommerce-jamstack/.

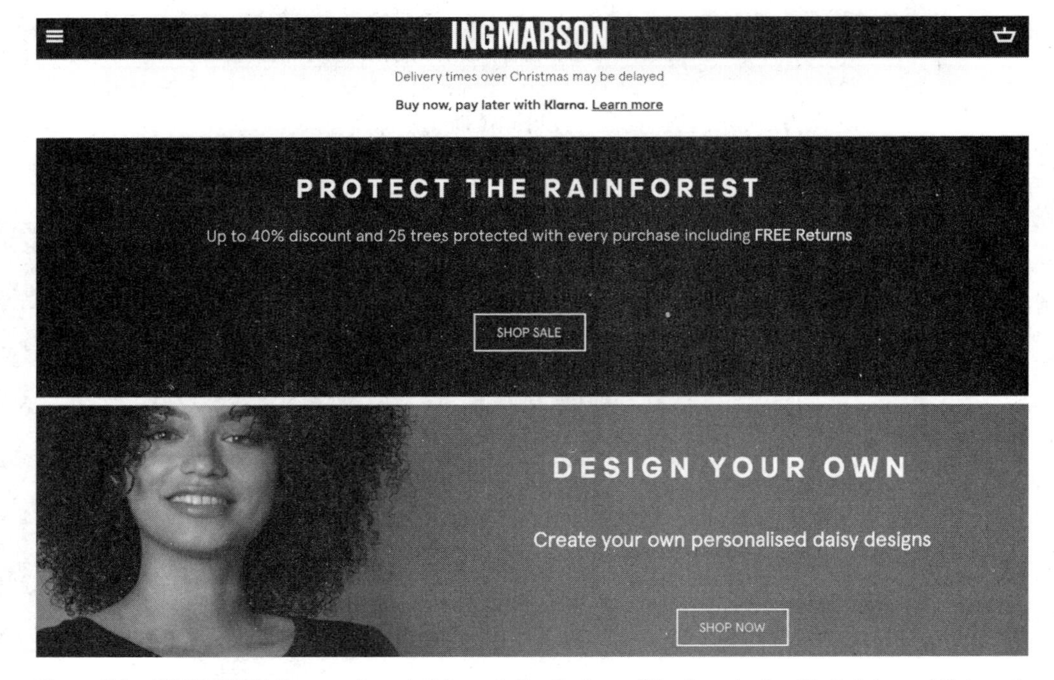

Figure 5.3 INGMARSON is an online clothing retailer that used the Jamstack with Gatsby and Snipcart to enable its e-commerce features.

COMMERCE.JS

Commerce.js (https://commercejs.com/) is an example of an API-based headless e-commerce solution that has been used to build sites like INGMARSON, shown in figure 5.3. It provides all of the tools for building an e-commerce site—product management, discounts, shopping cart, checkout, and more—that can be managed via a web-based backend but are accessed within your application through the JavaScript SDK and command-line tool. To help you get started, they also provide prebuilt example sites using popular web frameworks like React and Vue. Our sample application will use Commerce.js.

5.2.3 Why Commerce.js?

The decision to use commerce.js wasn't purely driven by the technical requirements of our sample application. The goal of the sample application, beyond being functional, is to teach some of the underlying concepts required to use Jamstack tools and frameworks. By requiring more custom code, using an API-based solution will allow us to explore more of the code needed to build a Jamstack application.

If you are evaluating which type of solution to choose, here are my recommendations:

- Choose an add-on solution if your priority is building your site quickly and easily over customizing the shopping cart or checkout process.

- Choose an API-based solution if your priority is to maintain control over the design and user experience, including the shopping cart and checkout, and you are comfortable with the additional code it will require.
- From a technical Jamstack point of view, you'll choose an all-in-one solution for the same reasons you choose an API-based solution. However, in some cases there may be additional features offered or existing business relationships that may make an all-in-one option a better fit.

5.2.4 Static site generator options

There aren't any specific static site generators (SSGs)—that I am aware of—that are geared toward e-commerce, so really any SSG will work. Nonetheless, we're planning on integrating with APIs both at build time—to populate product listings—and on the client-side—to enable shopping cart functionality. JavaScript-based SSGs make it easy to integrate with APIs during build time, and, more specifically, the JavaScript framework-based options can offer tools that make some of the client-side scripting easier. Let's look at a few of these options:

- *Gatsby*—Gatsby is a very popular React-based SSG. Some things that make Gatsby unique are its use of GraphQL for accessing data, including things like content and other internal data structures, as well as its plug-in system. Gatsby's large community has created thousands of plug-ins (over 2,500 as of this writing) that cover nearly any functionality or integration you might need.
- *Next.js*—Next.js is a React-based meta framework (a framework on top of a framework). It is not purely a static site generator. Next.js can be used for creating a standard React single-page application (SPA) with server-side rendering (SSR). It also provides tools to generate sites as static assets for a Jamstack application, but it even allows a site to determine whether a specific route (i.e., a path within the application) should be dynamic or static. This allows developers to build applications that are "hybrid," combining both SSR and static.
- *Nuxt.js*—As the name might imply, Nuxt.js shares a lot of similarities with Next.js, including the ability to use it for either SSR or static (and Nuxt 3 has added support for the hybrid SSR/SSG approach). However, Nuxt.js uses the Vue framework rather than React.
- *Gridsome*—Gridsome has a lot in common with Gatsby. It includes a lot of core features that define Gatsby—its use of GraphQL, its focus on generating static assets for Jamstack, and its plug-in ecosystem—but uses Vue instead of React.
- *Scully*—While there are numerous React-based and Vue-based SSGs, Scully is, as of this writing, the only option for developers who prefer to use the Angular framework. It is geared toward developing pure static-based Jamstack applications and has a plug-in ecosystem like Gatsby and Gridsome.

We'll be using Next.js to build our sample application.

5.2.5 Why Next.js?

So which static site generator should you use? Honestly, it mostly comes down to personal preference. Unless your e-commerce site has a specific need that is addressed by the availability of SSR in either Next.js or Nuxt.js, any of the options listed will work just as well. The question then becomes: do you prefer React, Vue, or Angular? Beyond that, it is simply a matter of features (GraphQL, plug-ins) or style preference.

Recent versions of Next.js also offer a new form of rendering called *incremental static regeneration* (ISR) that can be particularly useful for e-commerce sites with a large number of products and therefore lots of pages. ISR essentially defers the rendering of a page until it is first requested by a user. This means that an e-commerce site with thousands of products might only need to generate the most popular 200 product pages and render the remaining pages when they are first requested. This means that the first user who requests a page may see a small delay in receiving the page, but subsequent users will receive the page as if they were statically generated. You can learn more about ISR in the Vercel docs (http://mng.bz/lanB).

5.3 Getting set up to build the example e-commerce site

Now that we've made our tool choices—Commerce.js for e-commerce and Next.js for the SSG—let's get started building the sample application. The first thing we need to do is get everything set so that we can begin coding.

5.3.1 Setting up Next.js

Next.js doesn't install like some of the examples in prior chapters that used binaries or global npm installs. Instead, Next.js provides a tool called Create Next App (https://nextjs.org/docs/api-reference/create-next-app) to generate a new site, either a blank or one from a long list of starter templates. This doesn't require you install anything but instead a `npx` command: a package runner built into npm that can run an npm script without requiring an install. You can run `npx create-next-app` without any parameters, and it will create a simple one-page web app with the basic default Next.js application files and folder structure.

Create Next App can also use a template. This can be any GitHub repo, but the Next.js team already provides over 250 examples (https://github.com/vercel/next.js/tree/main/examples) that you can use. We'll use one of these examples, with-tailwindcss, as the basis for our example e-commerce store that sells toy figurines. We're going to build a product listing page, product detail page, and shopping cart, but to make this easier, we'll take advantage of the starter templates that Next.js provides. The particular starter we've chosen includes the Tailwind CSS library that will give us some basic styling building blocks to work with. This is a huge help: while I am many things, a designer is not one of them. To learn more about Tailwind, check out tailwindcss.com.

Let's get started by running `create-next-app` and specifying the Tailwind CSS example. Run the following command wherever you keep your web projects:

```
npx create-next-app -e with-tailwindcss
```

Create Next App will ask you for a project name. This will be used as the name of the folder that the project will be placed within. For this example, let's use next-ecommerce. Create Next App will not only generate the project files, but also install all the dependencies for us. All we need to do is change the directory into the project folder and run it:

```
cd next-ecommerce
yarn dev
```

If you prefer to use npm instead of Yarn, you can alternatively run `npm run dev`.

By default, the site will be running at http://localhost:3000. If you open that in the browser, you should see the standard Next.js start shown in figure 5.4.

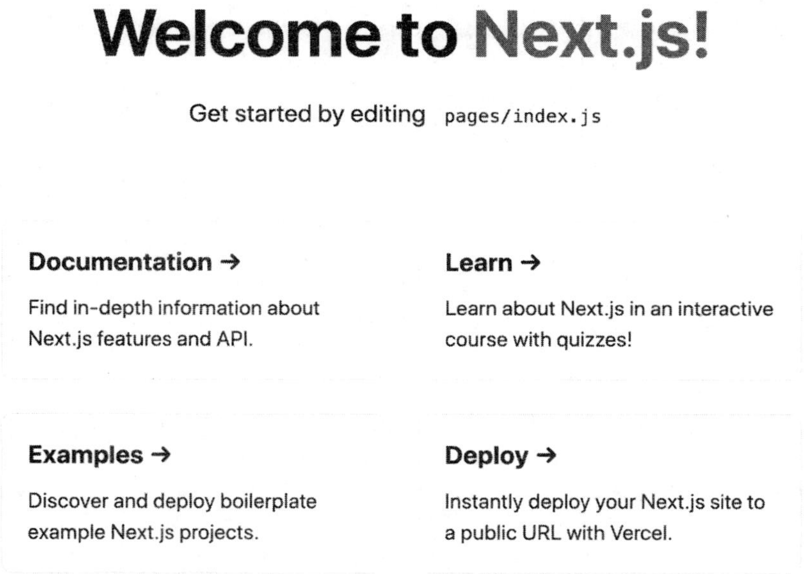

Welcome to Next.js!

Get started by editing `pages/index.js`

Documentation →

Find in-depth information about Next.js features and API.

Learn →

Learn about Next.js in an interactive course with quizzes!

Examples →

Discover and deploy boilerplate example Next.js projects.

Deploy →

Instantly deploy your Next.js site to a public URL with Vercel.

Figure 5.4 The default Next.js site that is generated when running `npx create-next-app`

Let's start by adding a basic navigation component for this site (listing 5.1). The pages we'll link to aren't created yet, but we'll fix that soon enough. First, create a components folder in the root of your site and create a file named nav.js inside that folder. We'll have two navigation items that are set in the `links` variable as "My Cart" and "About." These will link to `/cart` and `/about`, respectively. We loop through the links in this array using the JavaScript `map()` function to create the navigation.

Listing 5.1 The updated navigation in /components/nav.js

```
import Link from 'next/link';

const links = [
  { href: '/cart', label: 'My Cart' },
  { href: '/about', label: 'About' },
];

export default function Nav() {
  return (
    <nav>
      <ul className="flex items-center justify-between p-8">
        <li>
          <Link href="/">
            <a className="text-blue-500 no-underline text-accent-1 dark:text-
            ➥ blue-300">
              Jam Store
            </a>
          </Link>
        </li>
        <ul className="flex items-center justify-between space-x-4">
          {links.map(({ href, label }) => (
            <li key={`${href}${label}`}>
              <Link href={href}>
                <a className=" no-underline px-4 py-2 font-bold text-white
                ➥ bg-blue-500 rounded">{label}</a>
              </Link>
            </li>
          ))}
        </ul>
      </ul>
    </nav>
  );
}
```

We'll use this component to display the navigation across our pages. Let's start by clearing out much of the default content on the default site's home page and include the navigation component. For now, the page only imports the `nav.js` component and displays it above the main section that, for now, contains some dummy text.

Listing 5.2 Including the navigation in /pages/index.js

```
import Nav from '../components/nav';

export default function IndexPage({ products }) {
  return (
    <div>
      <Nav />
      <section className="text-gray-700 body-font">
        <div className="container px-5 py-24 mx-auto">
          <div className="flex flex-wrap -m-4">
            The product list will go here.
          </div>
        </div>
```

```
      </section>
    </div>
  );
}
```

Now that our project code is set, we need to set up and populate Commerce.js so that we can populate the data on the page.

5.3.2 *Setting up Commerce.js*

To get started, we'll need to sign up for a Commerce.js account via commercejs.com. Don't worry: the free account is generous enough for the purposes of this example. Once you're signed up, you'll be brought to the dashboard. This is where we'll create the categories and products that will populate our e-commerce store. I'll provide you some guidance on what to populate these with, but I should note that you are free to use whatever you choose; there is nothing in the code we'll write that requires that you use the products and categories I use.

In order for our e-commerce toy figurine store to work, it will need some products, each of which will be assigned to a category. We'll need to populate these within Commerce.js via their web-based backend.

Let's start by creating some categories. As you saw in figure 5.1, our store is comprised of some toys, figurines, and rubber ducks. Thus, we'll create two categories: Figures and Ducks. To create categories, click the Products navigation item on the left-hand side of the Commerce.js dashboard and then choose Categories. We'll need to provide a category name, and it will auto-fill a slug for us. We can use the default that it provides (figure 5.5).

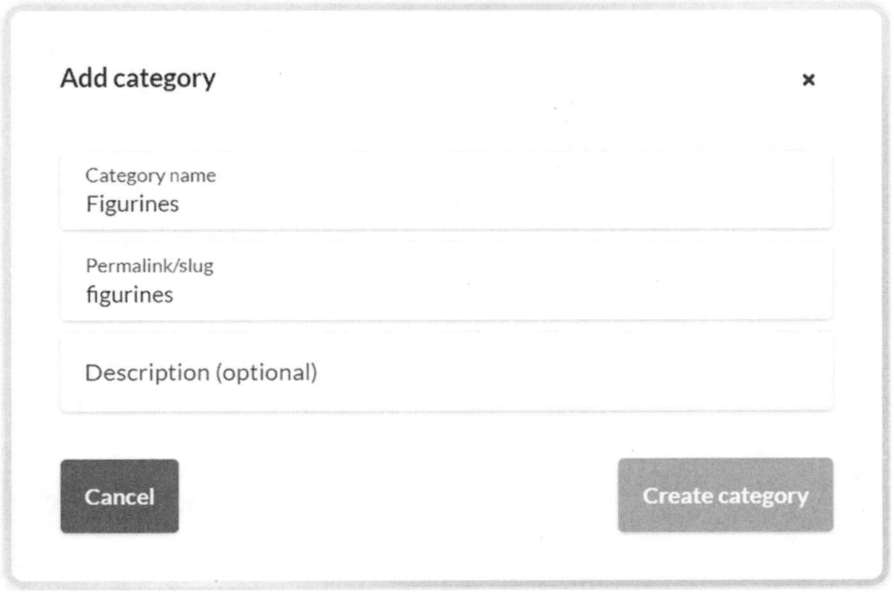

Figure 5.5 Create a new product category within the Commerce.js dashboard.

Next, let's create some products. Click the Products item on the left-hand navigation and then click Products. There is a lot of detail we can provide (see figure 5.6), but, for our purposes, we don't need to fill in all of them.

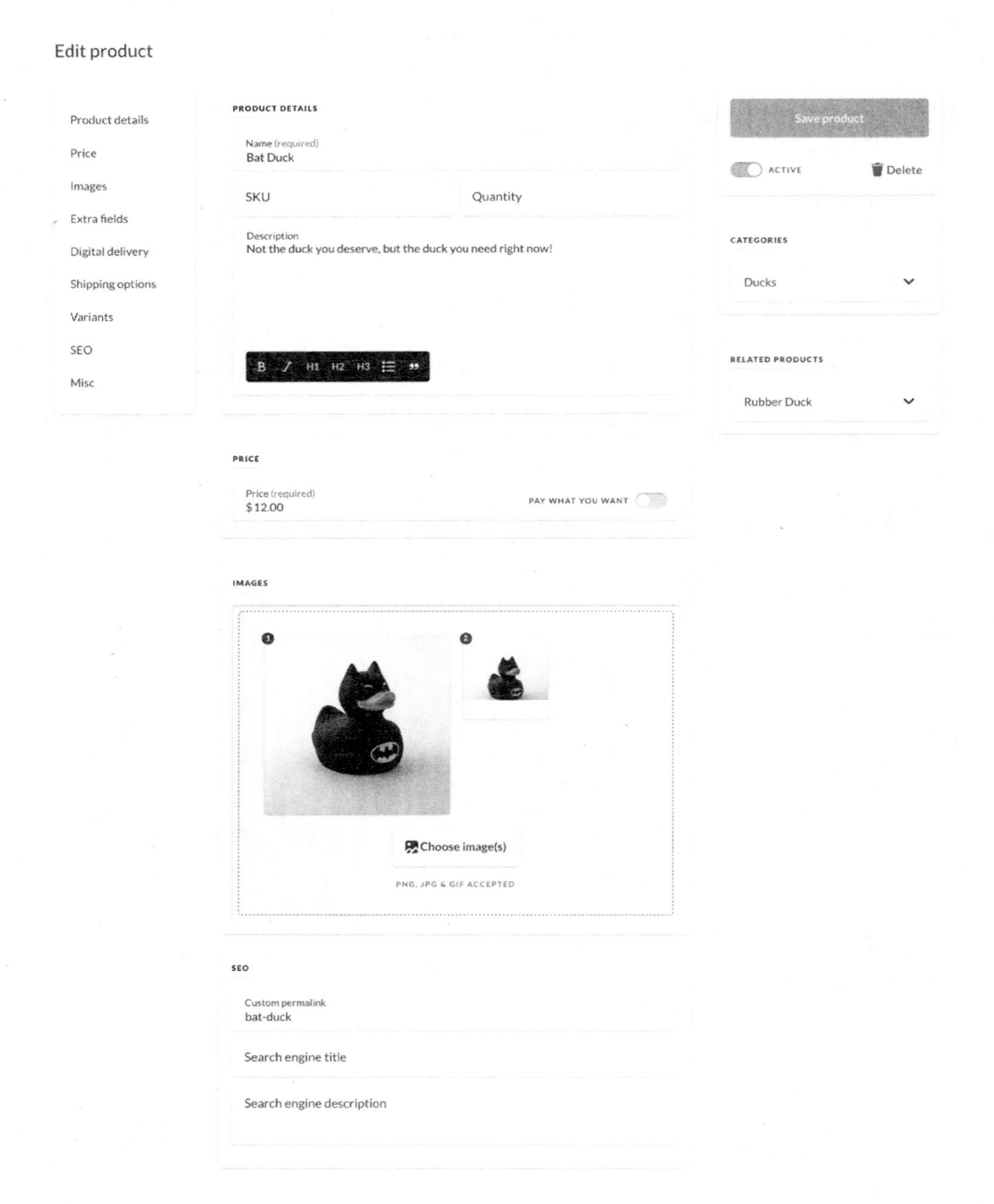

Figure 5.6 Create a new product within the Commerce.js dashboard. Please note that, for readability purposes, only the relevant sections of the page are displayed.

As we create new products, the key items to complete are as follows:

- *Name*—Feel free to name the product however you like.
- *Description*—Again, feel free to add whatever type of description you choose.
- *Categories*—Assuming you used the suggested categories, this will be either Figures or Ducks.
- *Price*—Provide whatever price you want.
- *Images*—I've provided some sample images in the book's GitHub repository (https://github.com/cfjedimaster/the-jamstack-book). For our design, we'll need two images for each product: one that is 400 × 400 px and one that is 350 × 192px.
- *Custom permalink*—A random permalink will be generated automatically if you don't provide one, but it's best that we do. This will be used within the application to construct the path to view the product.

Once we've created some products, we're ready to go back to the project we generated and start coding. But first, we'll need our API key to allow our project to access the data in Commerce.js. Click the Developer item on the left-hand navigation menu and then choose API keys. We'll just need to copy the public key.

5.3.3 Setting Next.js environment variables

If you still have the local web server running your site, go ahead and stop it for now, as we'll be making changes to the site's configuration.

In order to connect to the Commerce.js API from our site, we'll need the API key. However, we need to keep that key in a place that doesn't expose it directly in our code, which will be checked into GitHub. Next.js includes the ability to automatically load environment variables (see https://nextjs.org/docs/basic-features/environment-variables), which can be used to keep secrets like API keys or general configuration variables that you intend to reuse throughout your application. We will store our Commerce.js public API key as an environment variable.

Create a file named .env.local in the root of your project. By default, any variable we store in this file will be accessible via `process.env.ENV_VARIABLE_NAME`. For instance, a variable named `CHEC_PUBLIC_KEY` would be accessible as `process.env.CHEC_PUBLIC_KEY`. This is only accessible in the Node.js environment, either on the server in the case of an SSR application or in the build when using Next.js as an SSG. If you need to access this variable via client-side JavaScript, you can prefix the name of the variable with `NEXT_PUBLIC_`.

We'll need this variable to configure the Commerce.js SDK at build time, so we do not need it to be public. Let's put the following contents in .env.local, replacing `{{MY_API_KEY}}` with your Commerce.js public API key:

```
CHEC_PUBLIC_KEY={{MY_API_KEY}}
```

The .env file is useful for keeping things like API keys that you don't want to publish in your project repository. While this is intended to be a public key and will be accessible for people who inspect any API calls we make from the client, it's still advisable to keep it out of your published code. Plus, this gives us a single place to store the variable and reuse it wherever necessary within the application. The .env.local should already be in the .gitignore file that was generated with the project.

5.3.4 Loading the Commerce.js SDK

Commerce.js provides a Commerce.js SDK (https://commercejs.com/docs/) that helps make it easier for us to interact with the Commerce.js API via JavaScript. Let's install the SDK via Yarn. Run the following command. Be sure you are in the root of your site when running this command:

```
yarn add @chec/commerce.js
```

To configure the SDK, we need to pass the API key that we stored as an environment variable in the previous section. We don't want to have to pass this on every page that uses the SDK, so let's create a lib file that configures the SDK we can reuse throughout the application.

Create a folder named lib in the root of your project and then a file named commerce.js. This file will import the SDK and set the API key within a new instance of the `Commerce` object:

```
import Commerce from '@chec/commerce.js';

export default new Commerce(process.env.CHEC_PUBLIC_KEY);
```

Now, whenever we need to use the Commerce.js SDK, we only need to import `lib/commerce`, and we'll have access to the already configured Commerce.js SDK object.

5.4 Building the Jam Store e-commerce site

Everything is set up for our example e-commerce site that sells toy figurines. We've run `create-next-app` to generate a project with some basic site code for us to start with, which includes Tailwind CSS for some styling. We've also populated Commerce.js with some sample products and categories that will populate our store. Finally, we set up the Commerce.js SDK within our project and the API key stored as an environment variable so that we can connect to Commerce.js from our site. Now we can begin to create our product listing page, product detail page, and shopping cart to make our e-commerce site complete.

We're going to cover a lot of code in the upcoming sections. The goal here is to explore some of the key aspects you'll need to understand how to build sites using Next.js. It isn't critical that you understand every line of each code listing. I'll call out any of the critical concepts and portions that you need to pay attention to.

5.4.1 Creating the product listing component

Let's build the home page that will list all of the available products pulled from Commerce.js. We'll get some help from Tailblocks (https://tailblocks.cc/), a repository of ready-to-use code blocks built for Tailwind. Our product listing and product detail pages will be based on some of the code samples from their e-commerce category.

The product listing will loop over the products and display a product card. Let's create the product card as a reusable component. Create a folder named products within the components directory, and then create a file named ProductList.js inside the folder.

Let's look at the code for the ProductList.js component that will be used to generate each product card in the product listing. This component takes a `product` object that is passed via `props` and populates the values in the product card.

We'll write `ProductList` to accept properties (`props`) that we'll pass in when we use this component. This is how we'll pass in the product that will populate the page.

The component will return the HTML that will be rendered by the component, populating the various dynamic elements using the product details in the product property that is passed.

Listing 5.3 The product card component in /components/products/ProductList.js

```
import Link from 'next/link';

export default function ProductList({ ...props }) {
  const thumbnail = props.product.assets.filter((item, index) => {
    return (item.image_dimensions.width === 350);
  })[0];

  return (
    <div className="lg:w-1/4 md:w-1/2 p-4 w-full">
      <Link href={'/product/' + props.product.permalink}>
        <a className="block relative h-48 rounded overflow-hidden">
          <img
            alt={props.product.name}
            className="object-cover object-center w-full h-full block"
            src={thumbnail.url}
          />
        </a>
      </Link>
      <div className="mt-4">
        <h3 className="text-gray-500 text-xs tracking-widest title-font mb-1
        ➥ uppercase">
          {props.product.categories[0].name}
        </h3>
        <h2 className="text-gray-900 title-font text-lg font-medium">
          {props.product.name}
        </h2>
        <p className="mt-1">{props.product.price.formatted_with_symbol}</p>
      </div>
    </div>
  );
}
```

You may notice that we import the next/link component (https://nextjs.org/docs/api-reference/next/link). This is a helper component that is built into Next.js for client-side route transitions that are common in most single-page applications, enabling faster page loads. In this component, we use the `link` component to link to the product detail page using the `permalink` property we set on the product in Commerce.js.

One other thing worth noting is that each product returned by Commerce.js will have an array of images associated with it—in our case two because we've added two images for each product. For the product listing, we need the smaller thumbnail that is 350 pixels wide. In order to get this, we use a JavaScript array filter function (http://mng.bz/Bx8r) to loop through the product assets array and filter any images that are not 350 pixels wide. We set the `thumbnail` variable to the first item in the returned array.

5.4.2 Building the product listing

Let's combine the `ProductList` component and the commerce lib file to create the home page listing. Let's look at the final code for index.js. This code replaces the existing contents of index.js that we created earlier. This page calls the Commerce.js API within a special method of Next.js called `getStaticProps()`.

The `getStaticProps()` method is one of Next.js' built-in data-fetching methods. It is called at build time, so it is specifically designed for static routes within your Next.js application. Because we are building a typical Jamstack application, all of our routes are generated as static assets. We can use `getStaticProps()` to get any data our page needs and add it to the `props` object. In this case, we only need the array of products that will be passed into our output, where we will loop though the products, passing each product to the `ProductList` component we created earlier to output the product card.

Listing 5.4 The home page, including the product listing in /pages/index.js

```
import Nav from '../components/nav';
import commerce from '../lib/commerce';
import ProductList from '../components/products/ProductList';

export default function IndexPage({ products }) {
  return (
    <div>
      <Nav />
      <section className="text-gray-700 body-font">
        <div className="container px-5 py-24 mx-auto">
          <div className="flex flex-wrap -m-4">
            {products.map((product, index) => (
              <ProductList product={product} key={index} />
            ))}
          </div>
        </div>
      </section>
    </div>
```

```
  );
}

export async function getStaticProps() {
  const products = await commerce.products.list();

  return {
    props: {
      products: products.data,
    },
  };
}
```

It is worth noting that you need to import both /lib/commerce.js, which sets up our Commerce.js SDK to connect to the API, and /components/products/ProductList.js, which is the component for the product card that we created earlier. As we loop through, passing each product as a property into `ProductList`, we also set a unique `key` property. This helps React identify changes within the virtual DOM.

You may also notice that our `getStaticProps()` method is set as `async`. This is done so that we can use the JavaScript `await` operator. This allows us to reduce the amount of code needed by waiting for the result of an API call that returns a JavaScript `promise`, as the Commerce.js SDK does. Thus, our `products` constant waits for the `commerce.products.list()` method to return a result, preventing us from trying to return the value in the props before the API response is received.

If we restart the server and reload the page in our browser, we should now see the products we entered via the Commerce.js dashboard listed. It should look like the image we saw in figure 5.1.

5.4.3 Building the Product Detail page

We'll need to generate a product detail page dynamically for each product returned by Commerce.js (see listing 5.5). To do this, we're going to use a feature of Next.js called *dynamic routes* (https://nextjs.org/docs/routing/dynamic-routes). A dynamic route is recognizable because the file name is surrounded by square brackets. The text between the brackets will be the parameter we'll use to generate the page. Let's see how this works.

If you recall from when we created the `ProductList` component, we want the path to our product detail page to be /product/[permalink], where [permalink] is the permalink value we set in the product's properties in Commerce.js. We'll do this by creating a /pages/products folder and creating a [permalink].js file within that folder.

There will be a number of things happening on this page that we'll need to work out so that items can be added to the shopping cart, but, for now, let's focus on getting the detail pages to display. We render the product that we will pass to the component (we'll get to that in a moment), which pulls the proper image for the page based on the image width.

Listing 5.5 The Product Detail page in /pages/product/[permalink].js

```
import Nav from "../../components/nav";

export default function ProductDetail({ product }) {
  const fullImage = product.assets.filter((item, index) => {
    return item.image_dimensions.width === 400;
  })[0];

  return (
    <div>
      <Nav />

      <section className="text-gray-700 body-font overflow-hidden">
        <div className="container px-5 py-24 mx-auto">
          <div className="lg:w-4/5 mx-auto flex flex-wrap">
            <img
              alt="ecommerce"
              className="..."
              src={fullImage.url}
            />
            <div className="...">
              <h2 className="...">
                {product.categories[0].name}
              </h2>
              <h1 className="...">
                {product.name}
              </h1>
              <div
                className="leading-relaxed"
                dangerouslySetInnerHTML={{
                  __html: product.description,
                }}
              ></div>
              <div className="flex">
                <span className="...">
                  {product.price.formatted_with_symbol}
                </span>
                <button
                  className="..."
                >
                  Add to Cart
                </button>
              </div>
            </div>
          </div>
        </div>
      </section>
    </div>
  );
}
```

Perhaps you saw the strange property we used called `dangerouslySetInnerHTML`. Because the product description can contain HTML formatting, we need to render it using innerHTML in the DOM. However, in simple terms, since React uses a virtual DOM, we need to notify React that we are setting the innerHTML, which requires the use of the dangerouslySetInnerHTML method. (You can read this article [http://mng.bz/doYv] for a more in-depth explanation.)

At this point, all we've done is render the HTML, but, because this is a dynamic route, we need to tell Next.js what pages to render when it outputs the static files for our site. In our case, we plan on rendering one for each product using the permalink for each to determine the file name of the page. To do this, we'll need to add another special method built into Next.js for generating static sites called getStaticPaths(), as seen in listing 5.6.

Listing 5.6 The getStaticPaths() and getStaticProps() methods

```
export async function getStaticPaths() {
  const products = await commerce.products.list();

  // create paths with `permalink` param
  const paths = products.data.map((product) =>
     `/product/${product.permalink}`);
  return {
    paths,
    fallback: false,
  };
}

export async function getStaticProps({ ...ctx }) {
  const { permalink } = ctx.params;
  const product = await commerce.products.retrieve(permalink, {
    type: 'permalink ',
  });

  return {
    props: {
      product: product,
    },
  };
}
```

The getStaticPaths() method returns an array of path strings. We'll use Commerce.js to return a list of products and then populate these path strings with the product permalinks. The only other thing we need to return is a fallback key. When this key is false, as we've set, any path not returned by this method will return a 404. If it is true, Next.js will return a fallback version of the page instead of a 404. This would be useful if we were generating a large number of pages, causing a slow build. For any page not yet rendered, the user would see a loading indicator while getStaticProps() is called to populate the page.

Speaking of getStaticProps(), we need to create that function to populate the product details that will be rendered. In order to get the specific product details for this page, we'll use the `permalink` variable that is passed in via the page context variable (`ctx`). This is populated with the value from /product/[permalink] path that we generated using the product permalink. We can then use the permalink to query Commerce.js to give us the product detail that it matches.

Add both these methods to the [permalink].js page we created. You can place them below the code from listing 5.5.

The product detail we've created so far looks like figure 5.7.

Figure 5.7 The product detail listing for one of our products

5.4.4 *Enabling add-to-cart functionality*

So far, all the code we've written runs at build time, but each cart is unique to each user. Therefore, the add-to-cart functionality cannot be statically prerendered and needs to run on the client (i.e., the browser). To do this, we'll need to use the environment variable to configure the Commerce.js client-side script with our API key, which is currently only available at build time.

To make our `CHEC_PUBLIC_KEY` environment variable accessible to this script, we need to create a Next.js configuration file as next.config.js in the root of our project. In this file, we will tell Next.js to make this environment variable available on the client.

Listing 5.7 The Next.js configuration file in next.config.js

```
module.exports = {
  env: {
    CHEC_PUBLIC_KEY: process.env.CHEC_PUBLIC_KEY,
  }
};
```

Note that we could not use the `NEXT_PUBLIC_` shortcut that Next.js provides because Commerce.js expects the key using a specific variable name.

If you have your local server still running, you'll need to stop and restart it after making a configuration change.

Let's go back to our /pages/product/[permalink].js file and add some methods to enable the add-to-cart functionality. Because of our configuration change, we can now call Commerce.js from the client. Let's add a new method to the `ProductDetail` function in [permalink].js. Before we do that though, we need to import the Commerce.js library at the top of the file:

```
import commerce from "../../lib/commerce";
```

You can place the code for our `handleAddToCart()` function prior to the `return`. This method will tell Commerce.js to add the current product to our cart. For now, we'll just dump the complete cart contents into the browser console so that we can see that it's working:

```
const handleAddToCart = async (e) => {
  let cart = await commerce.cart.add(product.id, 1);
  console.log(cart);
};
```

Each time we call this method, we're telling Commerce.js to add one item of this product to the cart. This will either add a new product or increment the product if it already exists in the cart. Within the same file, let's call this method by modifying the Add to Cart button to call this function. Replace the current button code that is within the `render()` method with the following code:

```
<button
  onClick={handleAddToCart}
  className="..."
>
  Add to Cart
</button>
```

Once you save the file and it refreshes in your browser, open your browser developer tools console. When you click the Add to Cart button, you should receive something like the following response:

```
{success: true, event: "Cart.Item.Added", line_item_id:
➥ "item_7RyWOwmK5nEa2V", product_id: "prod_Op1YoV9x4wXLv9", product_name:
➥ "Bat Duck", …}
```

Of course, dumping results into the browser console isn't the ideal user interaction. What we need to do instead is let the user know that the item was successfully added.

We'll use a state variable to do this that will be set via React's `useState` hook. First, we need to add an import for the hook at the top of the file:

```
import { useState } from "react";
```

Now we can create the `cartText` state variable and the `setCartText()` function that will allow us to change the value of this state variable using this hook. State variables in React should not be changed directly, thus the need for the setter method. In addition, the `useState` hook allows us to pass in a default value, which, in this case, will be "Add to Cart." Add this line directly beneath the `export default function` line:

```
const [cartText, setCartText] = useState("Add to Cart");
```

Next, let's update the button again to use that variable instead of hardcoded text:

```
<button
  onClick={handleAddToCart}
  className="..."
>
  {cartText}
</button>
```

This by itself won't display anything differently than we were already displaying because we aren't updating the value on any state changes. To do that, we need to modify our `handleAddToCart()` method by removing the console log and adding some code to change the `cartText` state variable when the item is successfully added. We do not modify the state directly but instead use the `setCartText()` method to update it:

```
const handleAddToCart = async (e) => {
  let cart = await commerce.cart.add(product.id, 1);

  let cartText = "Added! (" + cart.quantity + ")";
  setCartText(cartText);
};
```

Now, if we try the page out, our cart text will display that the item was added and display the quantity of that item in the cart. Clicking the Add to Cart button should display "Added (1)" or, if the product was already in your cart, "Added (2)."

5.4.5 Building the shopping cart

At this point, users can view all of our products, click through to view specific products details, and add products to their cart. Next, let's allow them to view the items that are in their cart. To do that, we're going to borrow the design and layout from this pen in Codepen: https://codepen.io/abdelrhman/pen/BaNPVJO.

Create a /pages/cart.js file. This page will be different than prior pages, as it will not use `getStaticProps()`. Why? Because none of the properties that populate this page can be statically prerendered. The contents of the cart must be retrieved on the client, as it is tied to the specific user.

Instead of props, we'll make extensive use of state to populate the page with the user's cart details. This will allow us to update the state when the cart contents are retrieved and as the user interacts with the page to modify or remove items from their cart. Let's begin by creating the structure for our component and setting the state variables we'll need.

Listing 5.8 Creating the initial cart page in /pages/cart.js

```
import { useState, useEffect } from "react";
import Nav from "../components/nav";
import commerce from "../lib/commerce";

export default function ShoppingCart() {
  const [items, setItems] = useState([]);
  const [subtotal, setSubtotal] = useState({});
  const [total, setTotal] = useState({});
  const [checkoutURL, setCheckoutURL] = useState("");

  return (

  );
}
```

Next, let's populate the `return()`.The contents of our `return()`as it creates the shopping cart by looping through the items array that will contain our cart items and populating the remaining pieces of the output with other state variables are shown. Please note this leaves out some portions for readability purposes (denoted by a "..."). You can find the full code in the GitHub repository (http://mng.bz/8lRB).

Listing 5.9 The contents of the `render()` method in /pages/cart.js

```
return (
  <div>
    <Nav />
    <div className="container mx-auto mt-10">
      <div className="flex shadow-md my-10">
        <div className="w-3/4 bg-white px-10 py-10">
  ...
          {items.map((item, index) => (
            <div ... key={index}
            >
              <div className="flex w-2/5">
                <div className="w-20">
                  <img className="h-24" src={item.media.source} alt="" />
                </div>
```

```
              <div className="...">
                <span className="font-bold text-sm">
                  {item.product_name}
                </span>
                  <a href="#" ...>Remove</a>
              </div>
            </div>
            <div className="flex justify-center w-1/5">
              <button>...</button>
              <input className="..." type="text" value={item.quantity} />
              <button>...</button>
            </div>
            <span className="text-center w-1/5 font-semibold text-sm">
              {item.price.formatted_with_symbol}
            </span>
            <span className="text-center w-1/5 font-semibold text-sm">
              {item.line_total.formatted_with_symbol}
            </span>
          </div>
        ))}
      ...

      <div id="summary" className="w-1/4 px-8 py-10">
        <h1 className="...">Order Summary</h1>
        <div className="flex justify-between mt-10 mb-5">
          <span className="...">{items.length} items</span>
          <span className="font-semibold text-sm">
            {subtotal.formatted_with_symbol &&
              subtotal.formatted_with_symbol}
          </span>
        </div>
        <div className="border-t mt-8">
          <div ...>
            <span>Total cost</span>
            <span>
              {total.formatted_with_symbol && total.formatted_with_symbol}
            </span>
          </div>
          <button ...>Checkout</button>
        </div>
      </div>
    </div>
  </div>
</div>
);
```

We've set up the layout and the necessary state variables, but if you browse to the cart by clicking the My Cart navigation button, you'll see that the page has no contents—even if you have items in your cart (figure 5.8).

What we need to do next is load the contents of a user's cart from Commerce.js when the page is ready. For this, we'll use the `useEffect()` hook, which tells React to perform some action after the render is complete.

Figure 5.8 The statically prerendered cart page has no details. These must be loaded for each user via the browser.

Our `useEffect()` hook will call Commerce.js to get the user's cart and then update the component state variables to populate the page with the user's cart items and quantities. We've placed this code in a separate method contained within the hook because we need to call it asynchronously. (This is because asynchronous methods always return a promise, but `useEffect()` can only return a function, thus we cannot make `useEffect()` asynchronous.) Let's place this directly above the `return`:

```
useEffect(() => {
  async function fetchCart() {
    let cart = await commerce.cart.retrieve();
    console.log(cart);
    setItems(cart.line_items);
    setSubtotal(cart.subtotal);
    setTotal(cart.subtotal);
    setCheckoutURL(cart.hosted_checkout_url);
  }
  fetchCart();
}, []);
```

When the page reloads, you should now see the items you've added to your cart listed (figure 5.9).

So far so good, but the user is unable to add or subtract (and ultimately remove) items from the cart. Let's add a method right above our `useEffect()` method to handle that. This method will call Commerce.js to update the quantity and then update the relevant state variables:

```
const handleUpdateQuantity = async (id, quantity) => {
  let res = await commerce.cart.update(id, { quantity: quantity });
  let items = res.cart.line_items;
  setItems(items);
  setSubtotal(res.cart.subtotal);
  setTotal(res.cart.subtotal);
};
```

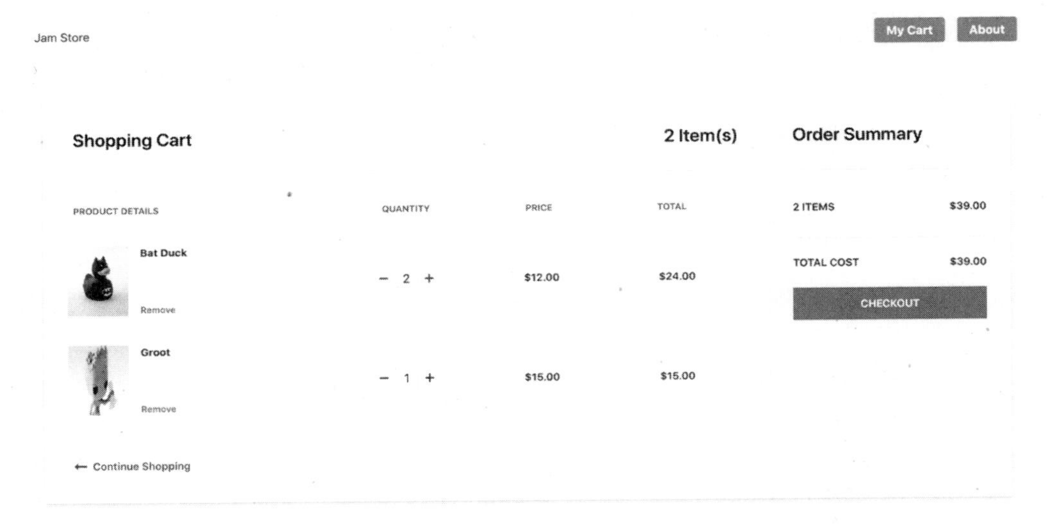

Figure 5.9 The contents of the cart are loaded on the client using the `useEffect()` hook and populated on the page.

Our `handleUpdateQuantity` method will handle adding, subtracting, or even manually supplying the quantity. First, call it from the Remove link. Removing it is simply a matter of setting the quantity to zero:

```
<a
  href="#"
  onClick={() => handleUpdateQuantity(item.id, 0)}
  className="..."
>
  Remove
</a>
```

Next, we'll add it to the subtract button by subtracting one from the current quantity:

```
<button
  onClick={() =>
    handleUpdateQuantity(item.id, item.quantity - 1)
  }
>
```

Finally, add it to the plus button by adding one to the current quantity:

```
<button
  onClick={() =>
    handleUpdateQuantity(item.id, item.quantity + 1)
  }
>
```

We have to handle the input field that allows a user to manually specify a new quantity differently. We need a method that will be called when the value is changed and, if the value provided is a number, update the quantity. This method can go under the `handleUpdateQuantity()` method:

```
const handleQuantityChange = (id, e) => {
  const quant = parseInt(e.target.value.trim());
  if (!isNaN(quant)) handleUpdateQuantity(id, e.target.value.trim());
};
```

We'll need to call the method when the user changes the value in the text box by adding an `onChange` event handler to the input field:

```
<input
    className="mx-2 border text-center w-8"
    type="text"
    value={item.quantity}
    onChange={(e) => handleQuantityChange(item.id, e)}
/>
```

You should now be able to increase, decrease, remove, and update the inventory of any item in your cart.

The last thing we need to do is enable the checkout process. Commerce.js offers the ability to use a hosted checkout option. This means that when a user clicks the Checkout button, they will be guided through the process on Commerce.js rather than through a customized checkout. Of course, should you want to, you can create a custom checkout, but, for the sake of simplicity, we're going to use the hosted option.

To do this, all we need is one simple handler function that we can place under the `handleQuantityChange()` method. This method simply opens a new window using the checkout URL that Commerce.js provides:

```
const handleCheckout = () => {
  // for now we're just opening a new window to the hosted checkout
  window.open(checkoutURL);
};
```

Then we need add the `click` event to the Checkout button so that it triggers that method when the button is clicked:

```
<button
  onClick={handleCheckout}
  className="..."
>
    Checkout
</button>
```

Clicking the Checkout button will now open a new tab with the hosted checkout on Commerce.js (figure 5.10).

Figure 5.10 Our e-commerce store uses a hosted checkout on Commerce.js rather than a customized checkout.

5.4.6 Adding Markdown content

Our e-commerce experience is complete, but we still have one final touch to put on our site: the About page that will be driven by file-based Markdown content. Next.js does not have Markdown support built in, but it's relatively easy to add. Let's see how.

We're going to need three npm plug-ins to enable this support:

- `raw-loader`—This package will enable us to import the raw Markdown files as strings within our Webpack configuration.
- `gray-matter`—Our Markdown files will contain front matter metadata. This library will allow us to easily parse that metadata.

- `react-markdown`—As we've discussed, React applications use a virtual DOM, and this library will render the Markdown within React's virtual DOM, meaning that React will be able to properly update only the changed DOM elements.

If you're still running your local site, you'll need to stop it first. Next, install all three libraries:

```
yarn add raw-loader gray-matter react-markdown
```

Let's use `raw-loader` first. To do this, we need to edit the Webpack configuration of our Next.js site. Webpack is a popular module bundler for building web apps that Next.js uses to bundle its JavaScript files for the browser. To edit the Webpack configuration, open next.config.js and add a new rule that looks for files with the .md extension and loads them with the `raw-loader`.

Listing 5.10 The Next.js configuration file at next.config.js to load Markdown files

```
module.exports = {
  env: {
    CHEC_PUBLIC_KEY: process.env.CHEC_PUBLIC_KEY,
  },
  webpack: function (config) {
    config.module.rules.push({
      test: /\.md$/,
      use: 'raw-loader',
    });
    return config;
  },
};
```

Create a new file at /pages/about.js for our About page. Within the `getStaticProps()` method, we can use the filesystem to load the raw Markdown file from /content/about.md. We'll then use the `gray-matter` library to read raw Markdown and separate the front matter metadata from the content. We'll pass the front matter and Markdown content as props to the page, where `ReactMarkdown` is used to render the Markdown as React components.

Listing 5.11 The About page that loads Markdown content at /pages/about.js

```
import fs from 'fs';
import Nav from '../components/nav';
import matter from 'gray-matter';
import ReactMarkdown from 'react-markdown';

export default function About({ frontmatter, content }) {
  return (
    <div>
      <Nav />
      <div className="content container px-5 py-24 mx-auto">
```

```
        <h1>{frontmatter.title}</h1>
        <ReactMarkdown children={content} />
      </div>
    </div>
  );
}

export async function getStaticProps() {
  const file = fs.readFileSync(`${process.cwd()}/content/about.md`, 'utf8');
  const data = matter(file);

  return {
    props: {
      frontmatter: data.data,
      content: data.content,
    },
  };
}
```

As you may have guessed from the code, before this page will work, we need to create a Markdown file at /content/about.md. Placing content files in a /content folder is a typical structure of Next.js sites that load Markdown content. My about.md file is pretty simple, but feel free to experiment by adding more Markdown markup of your own:

```
---
title: About the Jam Store
---

The Jam Store is built with:
* Next.js
* Commerce.js.
```

Let's restart our local site using `yarn dev` and see our About page by clicking on the About navigation item (figure 5.11).

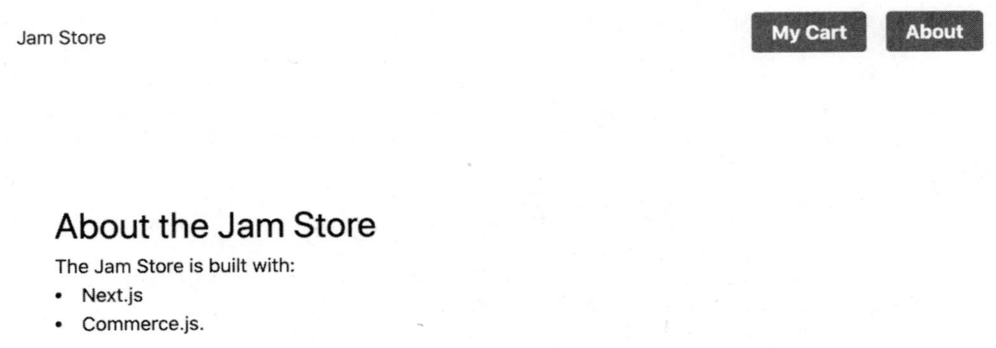

Jam Store

My Cart About

About the Jam Store

The Jam Store is built with:

- Next.js
- Commerce.js.

Figure 5.11 Our About page renders a title and body of the page from a combination of Markdown front matter and Markdown markup in the about.md file.

5.5 What's next?

We now have a fully functional e-commerce site, but there are a number of ways we can continue to improve it. The first and probably most obvious is to build the checkout process as custom rather than utilize the hosted checkout. We also never tackled issues such as discounts, sales tax, and shipping. These are things that can be managed and customized via the Commerce.js dashboard and then integrated into the site.

An e-commerce site has a lot of potential complexity that we don't have space to cover here. We chose to use Commerce.js as an API-based headless CMS in part because it helps illustrate many of the requirements for building a site with Next.js. However, one of the biggest decisions you'll make when building your own Jamstack e-commerce site is how much to favor the granular customizability of an API-based tool like Commerce.js over an add-on solution such as Snipcart that is faster to implement. There's no correct answer; it all depends on the requirements of your site.

Summary

- The dynamic user interface requirements of an e-commerce site make a perfect fit for the JavaScript-framework-based SSGs. JavaScript frameworks like React come with tools that make it easier to dynamically update the DOM in the browser, which can help you build highly dynamic page components like a shopping cart.
- A headless e-commerce system provides the backend to a Jamstack e-commerce site. There are three types of headless e-commerce systems:
  - All-in-one solutions are typically used for building both the frontend and backend of an e-commerce site, but generally provide headless options.
  - An add-on solution is designed to be easy to implement by providing both the UI and management for the entire cart and the checkout process.
  - An API-based solution allows granular customization by accessing all of its data and management capabilities via an API, but it requires more code to implement.
- Next.js is a React-based meta framework that provides tools for building either server-side rendering or static prerendering—or even a combination of both. Next.js makes it relatively simple to integrate with external API-based data sources like Commerce.js using built-in functions like `getStaticPaths()` and `getStaticProps()`.
- Next.js doesn't have built-in support for loading Markdown content, but it can be achieved through the use of multiple npm libraries. The `raw-loader` library provides the ability to import raw text files via Webpack. The `gray-matter` library reads the front matter metadata from a Markdown file. Finally, the `react-markdown` component renders Markdown within the React virtual DOM.

6 Deployment

This chapter covers

- Options for hosting your Jamstack site
- Using basic web servers
- Considering cloud file storage providers
- Choosing options tailored for the Jamstack

Congratulations! You've adopted the Jamstack way and have discovered the joy of using a local static site generator to turn dynamic content into simple files. The next step is to get those files on the internet so that they can be an actual website as opposed to a set of bits on your device.

In this chapter, we're going to cover multiple different options for getting those files online and available to the world at large. We'll cover the pros and cons of each, so you can determine which option makes sense for you and your project.

6.1 Web servers—The tried-and-true way

The first and simplest solution for hosting your Jamstack content is the solution we've used since the beginning of the web: a simple web server. Web servers like Apache (http://httpd.apache.org/) and IIS (https://www.iis.net/) have been around for decades and have powered everything from the smallest fan site to the largest e-commerce sites on the internet.

Using these options means simply taking the result of your static site generator (HTML, CSS, and other related files) and copying them somewhere under the "web root" of your existing web server.

That's it. Nothing more. You could create pipelines to automate this and handle the copying for you, or you could do everything by hand and use an FTP client to upload your files to your production server. At the end of the day this is the simplest option and would work best if you have an existing website and server already set up and simply want to use your Jamstack site as part of an existing solution.

As a concrete example of this, you can imagine an existing web server that uses PHP to serve dynamic web pages to users for an e-commerce site selling products. PHP would handle the shopping aspect, checkout, and so forth. You could then use the Jamstack to handle the documentation for products sold on the site. These wouldn't need PHP and can be simple static files. Obviously the entire site could be Jamstack (and in fact, the previous chapter discussed doing e-commerce with the Jamstack), but it's also possible to have sites where multiple solutions are in use!

6.2 Cloud file storage providers

Another option for hosting Jamstack sites is to make use of a cloud file storage provider. This is a service that simply lets you store files on the cloud and makes them accessible via HTTP. Cloud file storage providers are an attractive solution for developers who need a place to store files without the worry of managing a disk drive. By providing (near) infinite space, a developer can simply push files up and not worry about a full drive. (As long as they can handle the price, of course.)

Many of these providers now support an option to host a set of files as a website. Let's take a look at some of these options.

6.2.1 Amazon S3

As one of the oldest and most mature services, Amazon S3 is probably the first thing developers think of when considering cloud-based file storage. Amazon S3 (https://aws.amazon.com/s3/) is a very flexible, very robust file storage solution with great pricing (https://aws.amazon.com/s3/pricing). Amazon also provides a pricing calculator (https://calculator.aws) that lets you estimate your costs.

While every site is different, I can provide an example from my own usage. I have slightly less than a gig of files on S3 with about 70,000 requests over a month's period. My bill for that comes to about 3 cents. Yes, 3 cents. In order to enable HTTPS for those requests, my price skyrockets to 73 cents. Again, you need to look at your site, traffic, size, and so forth before committing, but most likely your costs will be minimal. Also note that Amazon offers various different "free tier" (https://aws.amazon.com/free/) choices. At the time this book was written, Amazon was offering 5 gigs of storage for free for 12 months.

To begin, you need to create an Amazon AWS account. Once you have, you can sign in to the AWS Management Console (figure 6.1).

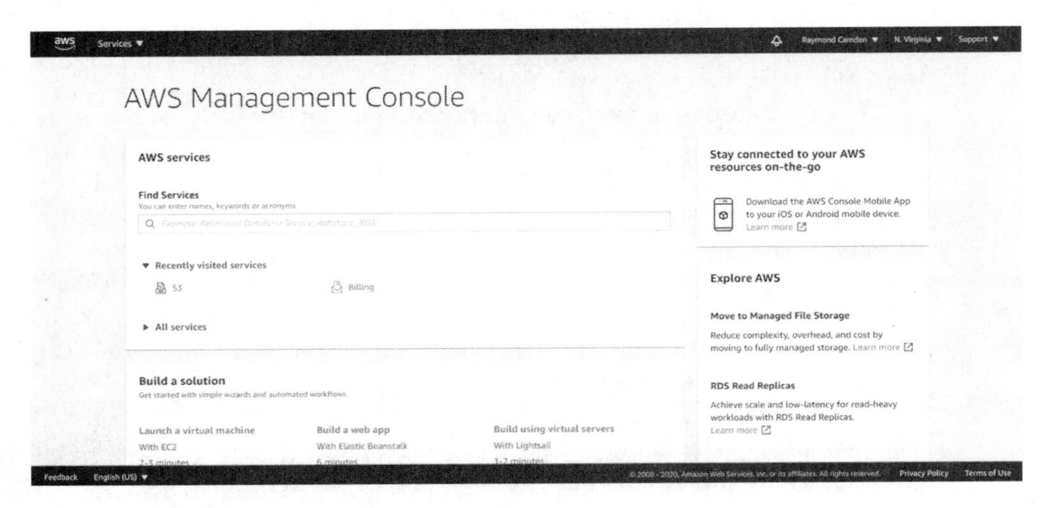

Figure 6.1 AWS dashboard

The console can be somewhat overwhelming, but you can use the Find Services search box to look for S3. Do that, and you'll end up on the S3 dashboard itself. Your dashboard may look a bit different based on whether you've used S3 before and whether Amazon has updated their UI (figure 6.2).

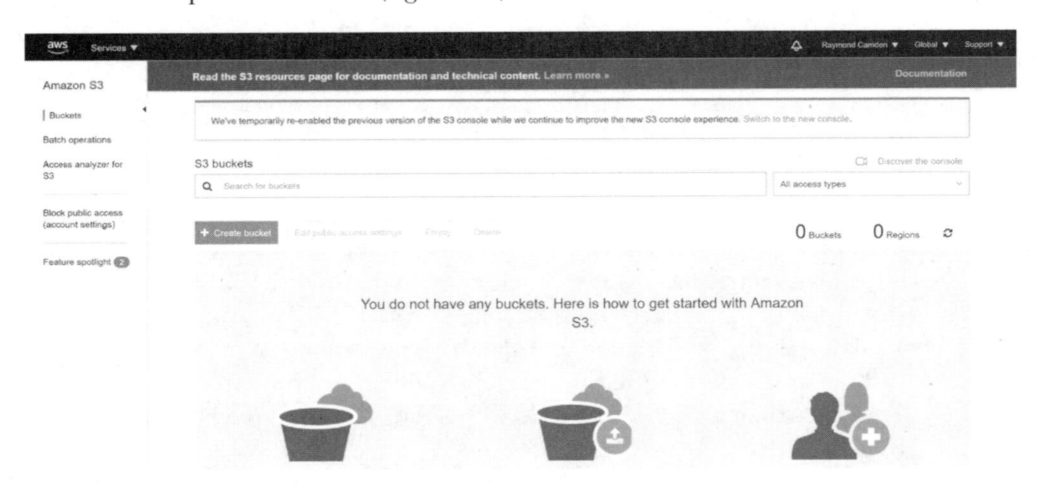

Figure 6.2 The S3 dashboard console. This will change after you've created your first bucket.

Amazon's S3 services lets you store files in *buckets*. You can think of a bucket as disk drives or folders for your project. While you can name your buckets anything you want, to host a website you need to name your bucket the same as your domain name.

For example, if I were hosting my blog on S3, I'd have to name my bucket raymondcamden.com.

> **NOTE** In order to support both raymondcamden.com and www.raymondcamden.com, you need to create *two* buckets. The bucket for www.raymondcamden.com would redirect to the first bucket. As we are going to be testing with a subdomain only, we'll only create the one bucket, but we will also show where that setting is so you can see how to do it yourself!

To begin, hit the Create bucket button. The name of the bucket should match the domain name of your site. In figure 6.3, I used a subdomain named jamstack.raymondcamden.com. You can leave the rest of the settings as they are.

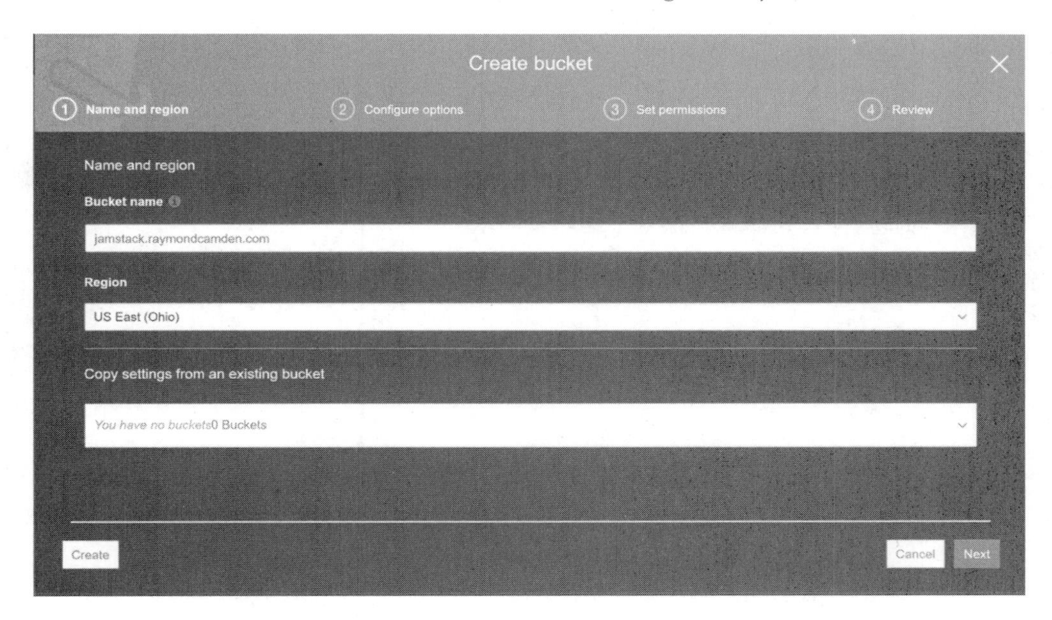

Figure 6.3 Defining the name of your new S3 bucket

Click Next through the next two steps and hit Create bucket when done. You'll then see your bucket listed (figure 6.4).

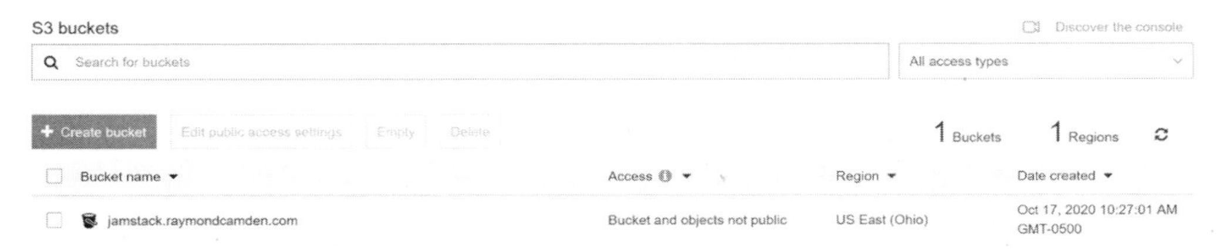

Figure 6.4 Your newly created S3 bucket

Select the checkbox by the bucket, and in the fly-out menu that appears, select Properties. This brings you to a set of options, including Static website hosting (figure 6.5).

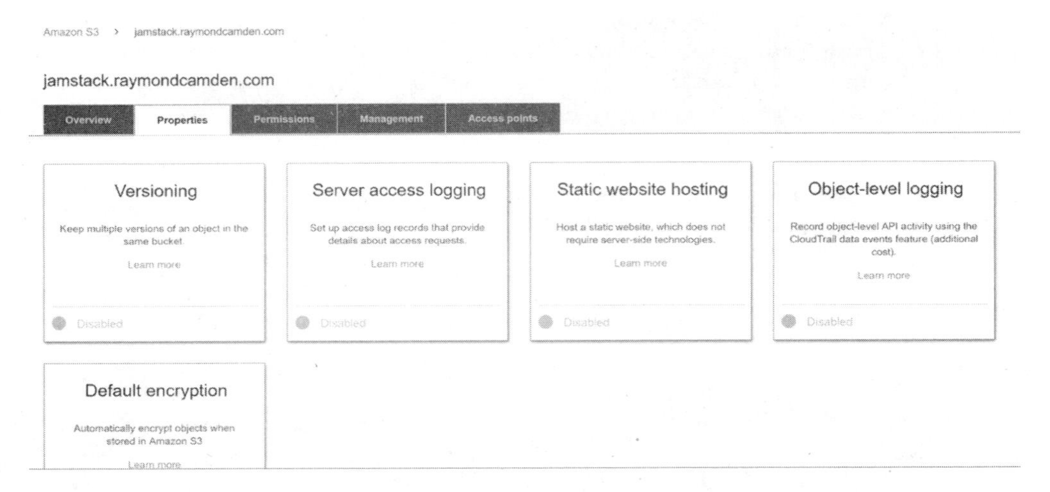

Figure 6.5 Various options for your S3 bucket

If you click that box, you get a new option for enabling website hosting, as shown in figure 6.6.

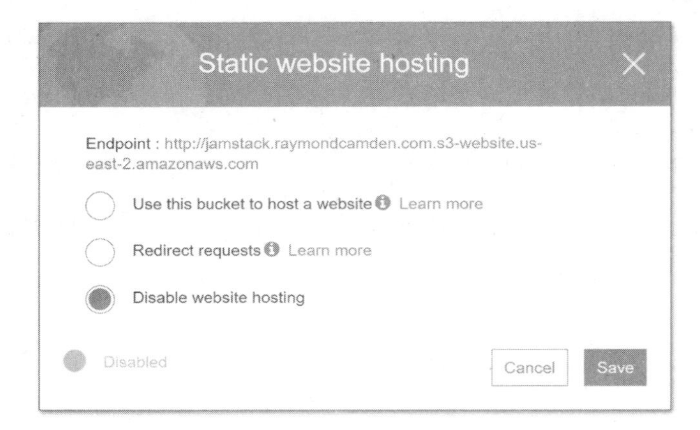

Figure 6.6 Dialog window to enable website hosting for a bucket

If you enable website hosting, you then need to specify what the index document is. This is the page loaded when a website is loaded and no specific file is requested. Almost always this should be index.html. Enter that value and you can then hit Save (figure 6.7).

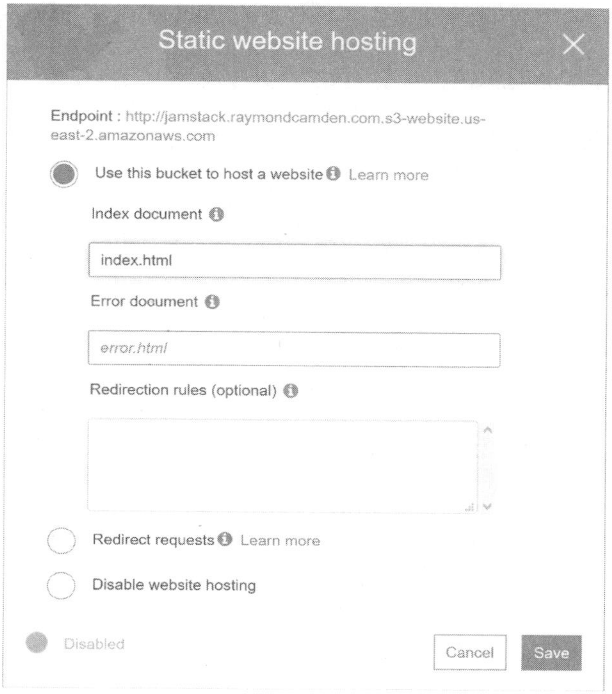

Static website hosting

Endpoint : http://jamstack.raymondcamden.com.s3-website.us-east-2.amazonaws.com

Use this bucket to host a website Learn more

Index document

index.html

Error document

error.html

Redirection rules (optional)

Redirect requests Learn more

Disable website hosting

Disabled

Cancel Save

Figure 6.7 Completing the options to enable website hosting

NOTE This is also where you would handle redirecting a www subdomain to your main bucket. We aren't covering that in this example, but now that you've seen where it's set, you can do this yourself if you need to.

After you have saved your changes, you can immediately hit your site using the URL shown in the dialog. If you didn't copy it, just click to open the Static website hosting option again. You should see something like figure 6.8.

403 Forbidden

- Code: AccessDenied
- Message: Access Denied
- RequestId: 9F560D4B1C008A88
- HostId: HhiXQaT8z16p+MZZInyGWa1B/mshbJeP+cM4PAwHz8gVYW4YwiX/LuCW7bi/FylKAJA10iFkQSQ=

Figure 6.8 403 error loading the new website

The reason you get this error is because you haven't actually uploaded a website yet! We're going to cover that in a moment, but before we do, we need to make another change.

By default, your bucket was created with all public access denied. This is a good and safe thing. However, before we start working with files, we need to edit the bucket so that the public has read access to it. This will then allow us to make our files public as well.

Select the Permissions tab in your bucket, and in the Block public access section, click the Edit button to turn off this setting (figure 6.9). This will give you a scary warning, but since we know what we are doing (allowing public access to static assets), it is safe to continue.

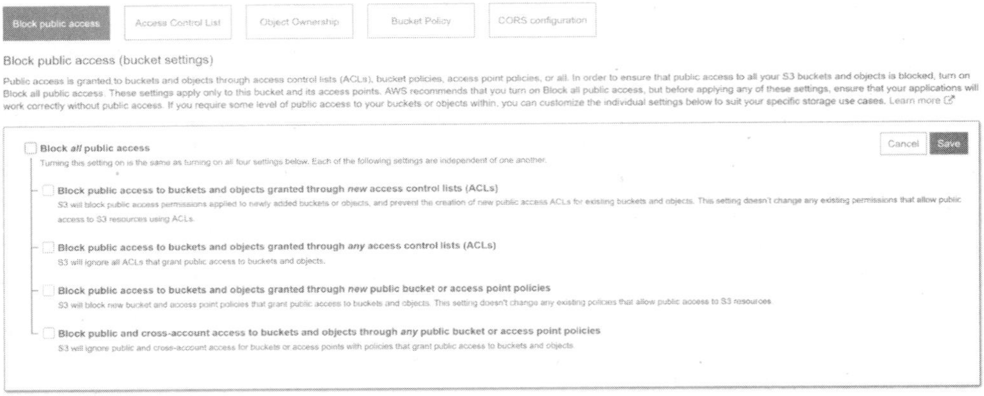

Figure 6.9 Disabling "Block all public access"

With that done, it's time to put things in the bucket. There are multiple ways you can put files into a S3 bucket; the simplest is to use the Upload option and simply drag your files and folders onto the browser. For smaller websites, this should be sufficient (figure 6.10). Remember, the UI may be different. Amazon constantly updates their UI and even shows different screens for different users. If what you see doesn't match exactly, do your best to find similarly named options.

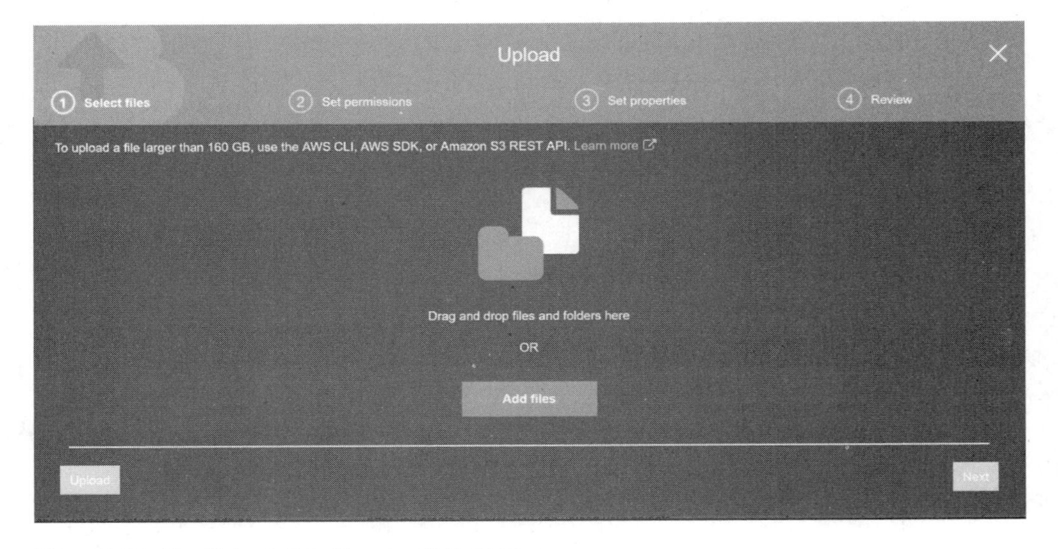

Figure 6.10 The Upload dialog for your S3 bucket

You can choose any of the previous examples in the book or any other website. For this demonstration, we will use the output from the Camden Grounds example in chapter 2. Remember that the Eleventy store's output is in the _site folder. You can select all the files and folders and drag them into the dialog. Before clicking Upload, click Next so that you can set permissions. At the bottom by "Manage public permissions," change the default to "Grant public read access to this object(s)." You can then click the Upload button (figure 6.11).

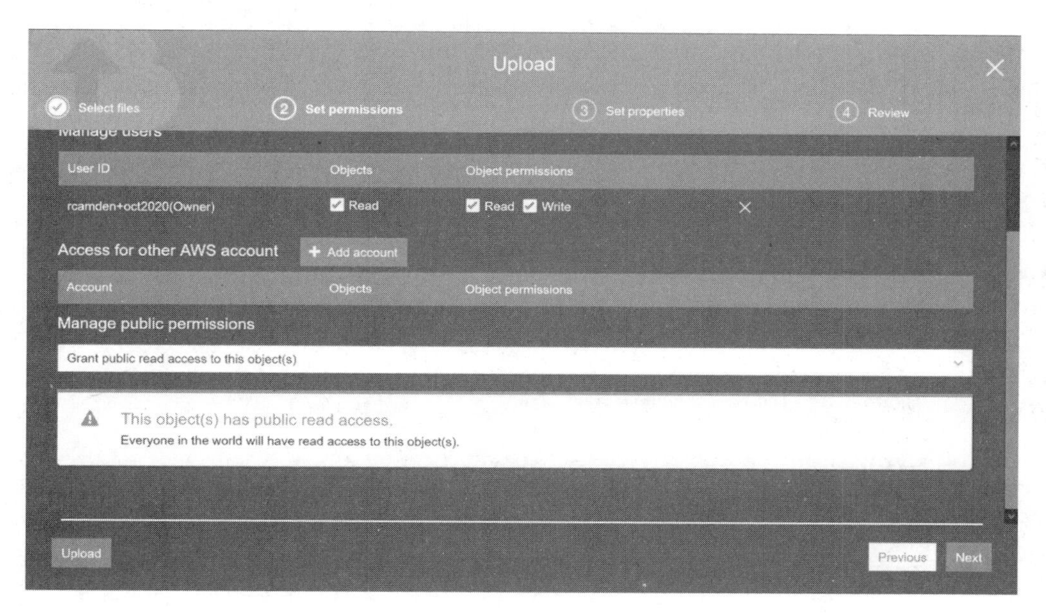

Figure 6.11 With permissions set, it's time to upload the files.

Depending on your bandwidth, and what you selected to upload, once you get confirmation that the upload is done, you can reload the earlier request and should now see your site (figure 6.12).

That's basically the process, but there's more you can do. For example, Amazon has a CDN service (CloudFront) that also enables you to use https. You can add redirects to your site as well. Now that we've looked at Amazon, let's look at some of your other options.

6.2.2 *Other cloud file storage hosting options*

While Amazon S3 is the best-known option for cloud-based file storage, it isn't the only option available. Another option to consider is Google Cloud Storage (https://cloud.google.com/storage). Like Amazon S3, Google Cloud Storage lets you store files of unlimited size with worldwide access. And, like S3 is a small part of AWS, Google Cloud Storage is but one aspect of the whole Google Cloud platform.

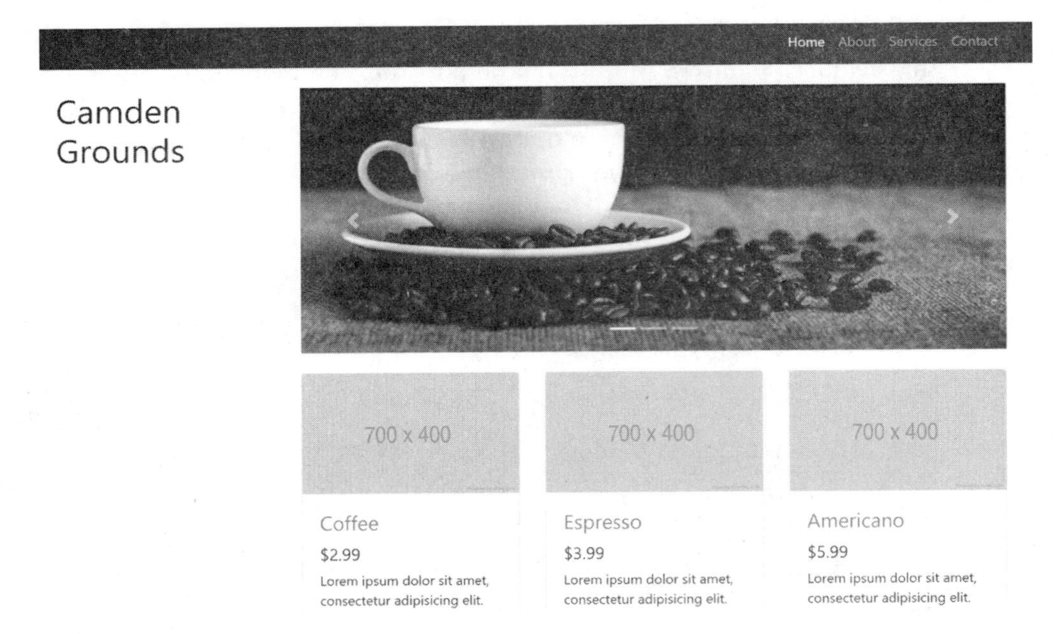

Figure 6.12 The S3-deployed website is live!

There is pricing information (https://cloud.google.com/storage#section-10) available, and like S3, you pay both for how much you store as well as how often the files are accessed. There is a free tier (http://mng.bz/Ex7j), which honestly was a bit hard to find at first, but would certainly let you test projects to see if you like the platform.

Google provides documentation (http://mng.bz/Nx77) for hosting a website, and it follows a similar pattern to Amazon's platform. Create a bucket, upload your files, and then share them publicly (figure 6.13).

Another cloud-based file service option is Azure Blob Storage (https://azure.microsoft.com/en-us/services/storage/blobs/). Its documentation for using it to host websites can be found here: http://mng.bz/Dx7a. Or, as an alternative, you might want to consider using a newer option from Microsoft, the Azure Static Web apps service (https://azure.microsoft.com/en-us/services/app-service/static/), which is specifically oriented toward Jamstack-style solutions.

You've now seen how a cloud-based file service can be an acceptable solution for hosting static websites. But, for the most part, that's all they do: show files. Let's look at services that are more tailored for the Jamstack.

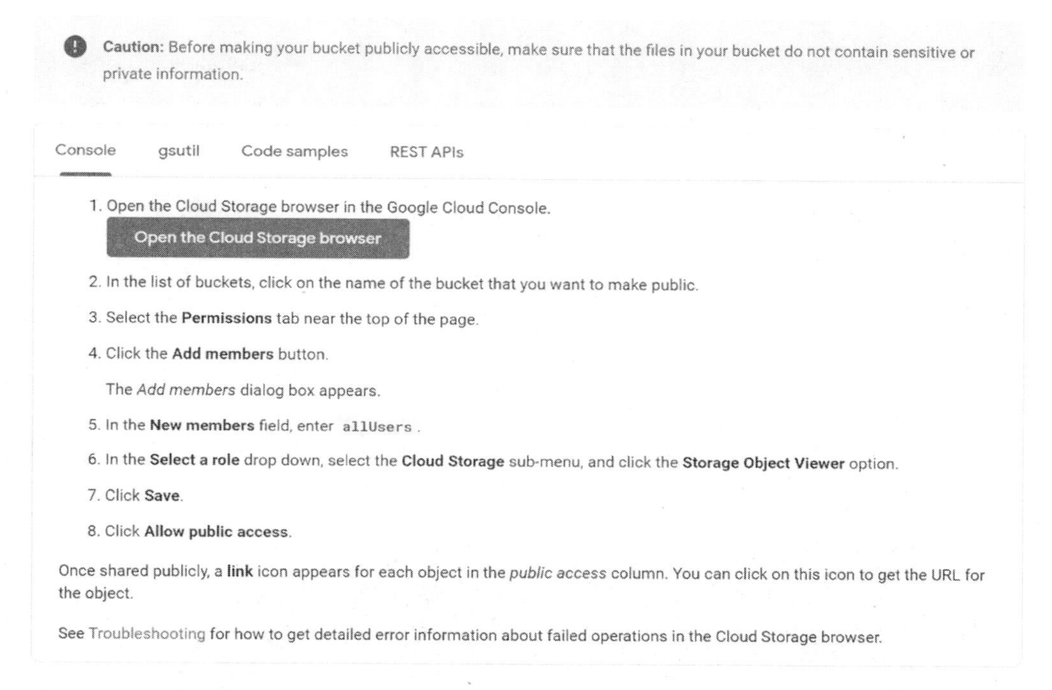

Sharing your files
To make all objects in your bucket readable to everyone on the public internet:
Caution: Before making your bucket publicly accessible, make sure that the files in your bucket do not contain sensitive or private information.
Console
gsutil
Code samples
REST APIs
1. Open the Cloud Storage browser in the Google Cloud Console.
Open the Cloud Storage browser
2. In the list of buckets, click on the name of the bucket that you want to make public.
3. Select the Permissions tab near the top of the page.
4. Click the Add members button.
The Add members dialog box appears.
5. In the New members field, enter allUsers .
6. In the Select a role drop down, select the Cloud Storage sub-menu, and click the Storage Object Viewer option.
7. Click Save.
8. Click Allow public access.
Once shared publicly, a link icon appears for each object in the public access column. You can click on this icon to get the URL for the object.
See Troubleshooting for how to get detailed error information about failed operations in the Cloud Storage browser.

Figure 6.13 Google Cloud instructions for sharing files

6.3 Azure Static Web Apps

Azure Static Web Apps (https://docs.microsoft.com/en-us/azure/static-web-apps/) is a new service by Microsoft on the Azure platform. It connects with GitHub repositories and uses GitHub Actions to automatically perform site updates on commits. It supports simple static files (e.g., the output of any of the generators previously covered in the book), as well as static site generators themselves.

For example, you can publish a Jekyll site and Azure Static Web Apps can be configured to perform Jekyll's build action to generate the final, static HTML. This (and the services that follow in this chapter) makes it more appropriate for the Jamstack. Finally, it also ties in well with Azure Functions, Microsoft's serverless platform.

Before looking at Azure Static Web Apps (or Azure in general), you will need a free Azure account. You can sign up (https://azure.microsoft.com/free/), and then go into your dashboard (https://portal.azure.com) and ensure you have at least one subscription. Think of a subscription as a high-level collection for things you use in Azure. After you've signed up, you can create a free tier subscription (figure 6.14).

Azure, like AWS, is both incredibly powerful but also complex. Luckily you can avoid the complexity by making use of the Visual Studio Code extension of the same

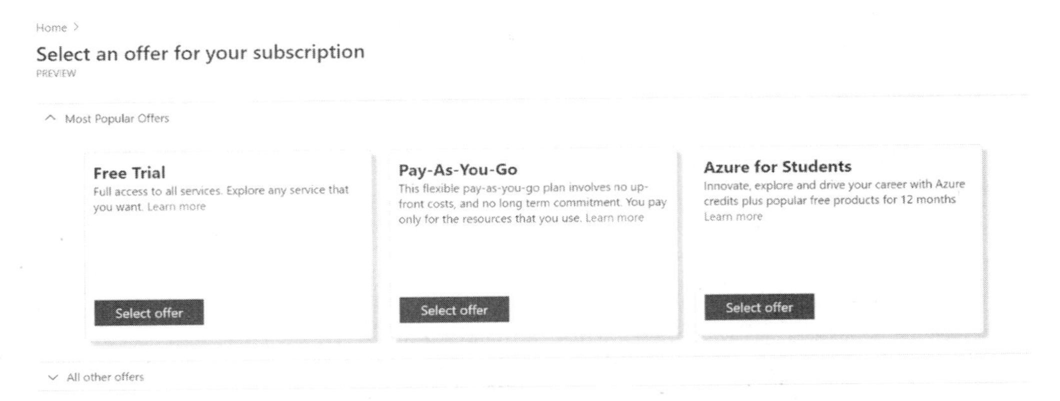

Figure 6.14 Azure interface to add a free subscription

name: Azure Static Web Apps. Visual Studio Code (https://code.visualstudio.com/) is a free, open source code editor from Microsoft. It's become incredibly popular recently with web developers. You can add this to your Visual Studio Code install by simply going to the Extensions panel and searching for it (figure 6.15).

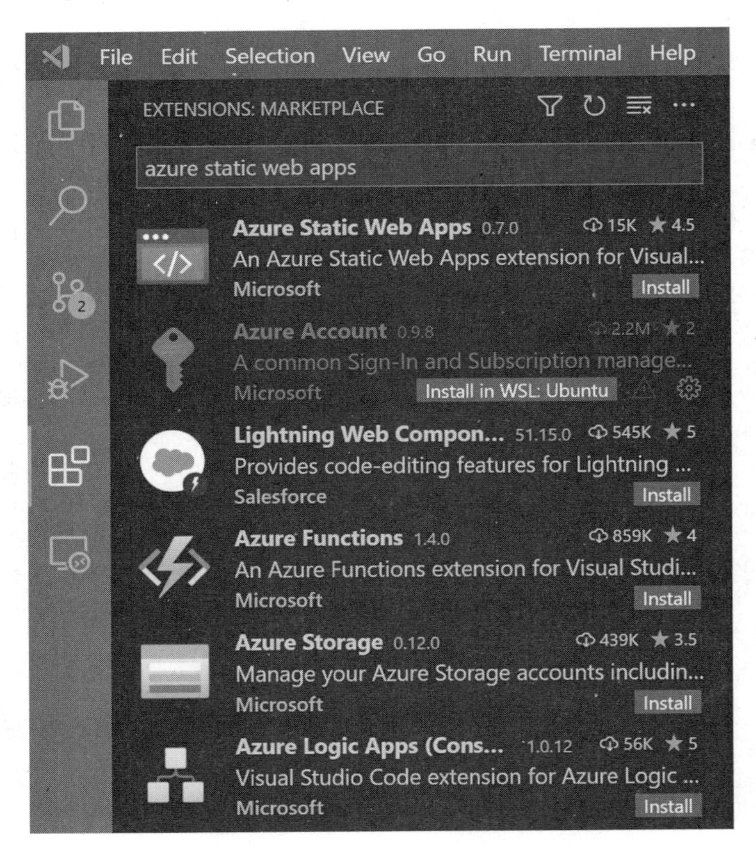

Figure 6.15 Discovering the extension via Visual Studio Code itself

After installing it, you'll see an A-shaped icon (the Azure logo, visible at the bottom in figure 6.15) that, once clicked, will open a new panel. The extension will walk you through the authentication process and have you select the subscription you created earlier.

To test this service, let's once again use the Eleventy site used for the Amazon S3 example. However, this time it's going to be a much more powerful integration. Instead of pushing the output of Eleventy to the S3 bucket, we'll use Azure Static Web App's GitHub integration to create a process where commits to the repository automatically fire off code to build the site and deploy to Azure.

To make things a bit simpler, the Camden Grounds site from chapter 2 has been copied to new GitHub repository at https://github.com/cfjedimaster/eleventy-for-azure. We cloned that to the local file system and made two changes.

First, we install Eleventy local to the project itself by running `npm install --save-dev @11ty/eleventy`. This updates the project's package.json file to list Eleventy as a development dependency. Next, we need to edit the `scripts` block to add a new script command that specifies how the site is built:

```
"scripts": {
    "start": "node_modules/.bin/gulp watch",
    "build":"npx @11ty/eleventy"
},
```

The start script was already there from the theme we used to build Camden Groups. The addition is the second line, starting with `build`. Azure is going to pick up on this when working with the repository and use it to build out the site. So how do we add this project to Azure?

In Visual Studio Code, open the folder containing the repository. Click on the Azure icon and in the Azure Static Web Apps panel, click the + symbol to create a new web app. You will first be prompted for a name. Enter CamdenGroundsAzure (figure 6.16). If asked to fork and clone the repo, do so.

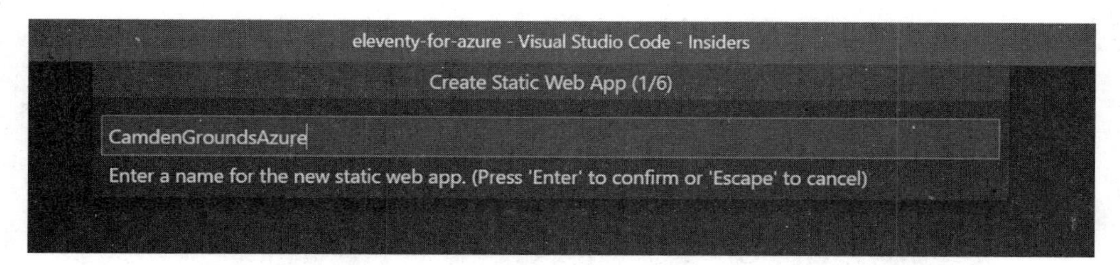

Figure 6.16 Naming the new application

The next prompt will be for the branch to use. GitHub is using main for the branch, but if you have an older project, your default branch may be master. It will show the branch in the dialog, so you can simply click on it.

Next, it will prompt you for the application code directory. Since the entire repository is the application, you will want to select the first option: / (figure 6.17).

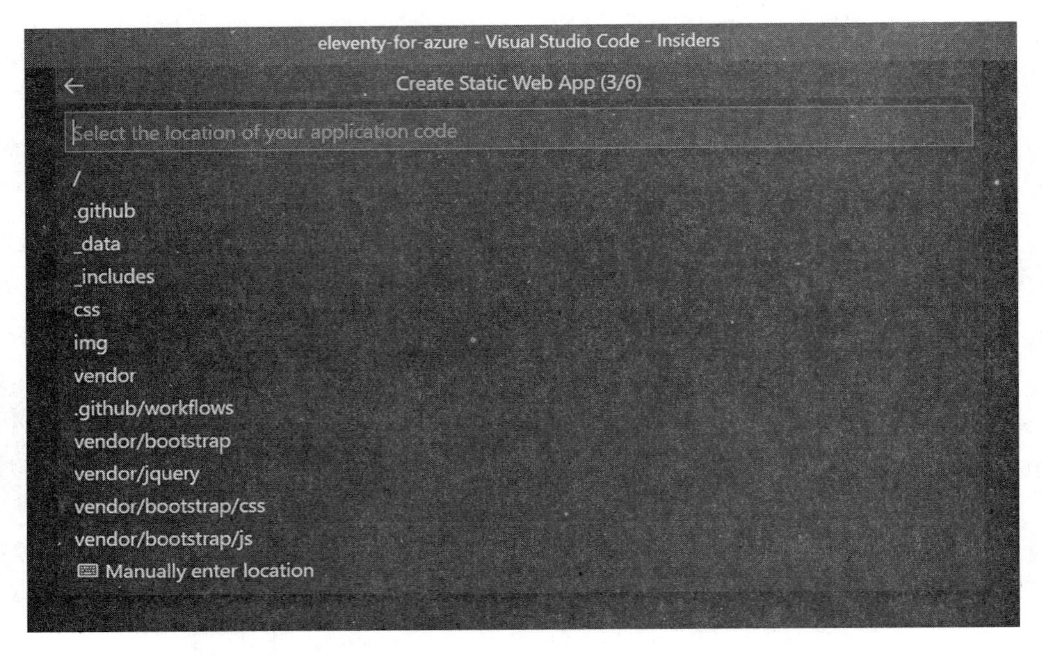

Figure 6.17 Selecting the application directory

The next prompt will ask you for the location of your Azure Functions code. We aren't using that, so select "Skip for Now."

The final prompt will be for the output path of your site: where your static site generator outputs the final HTML. For Eleventy, that's _site, so enter that in the dialog shown in figure 6.18.

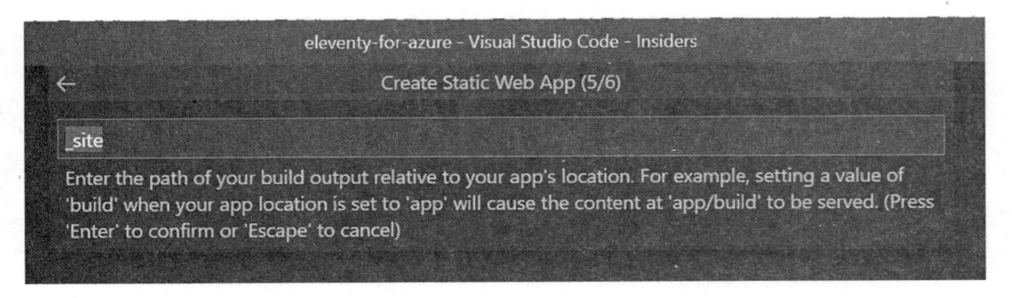

Figure 6.18 Entering the location where generated files can be found after the site is built

The final prompt refers to where Azure will deploy your site in its worldwide infrastructure. Take the default here, which will probably be central US.

At this point, the extension and Azure start doing a bit of work. They begin by adding the appropriate GitHub actions to your project (which you can see if you open the repository) that handle doing the push to Azure when files are committed. To see your project on Azure, click to open the free trial node in the Azure panel and you'll see your new project. Right-click on that, and select "Open in Portal" (figure 6.19).

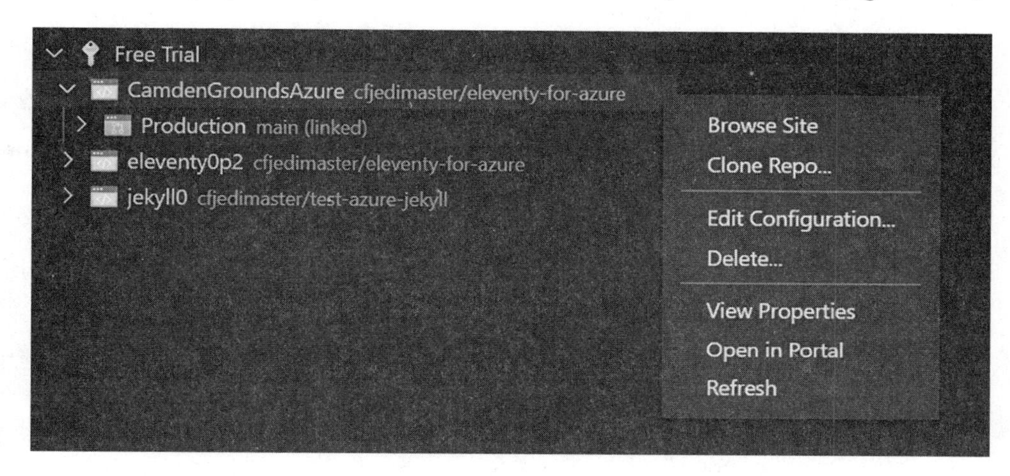

Figure 6.19 Right-click on your new project to see various options, including one to open the portal.

The portal contains a large amount of information, but what you want to focus on is the URL on the right-hand side, which you can use to test your site (figure 6.20).

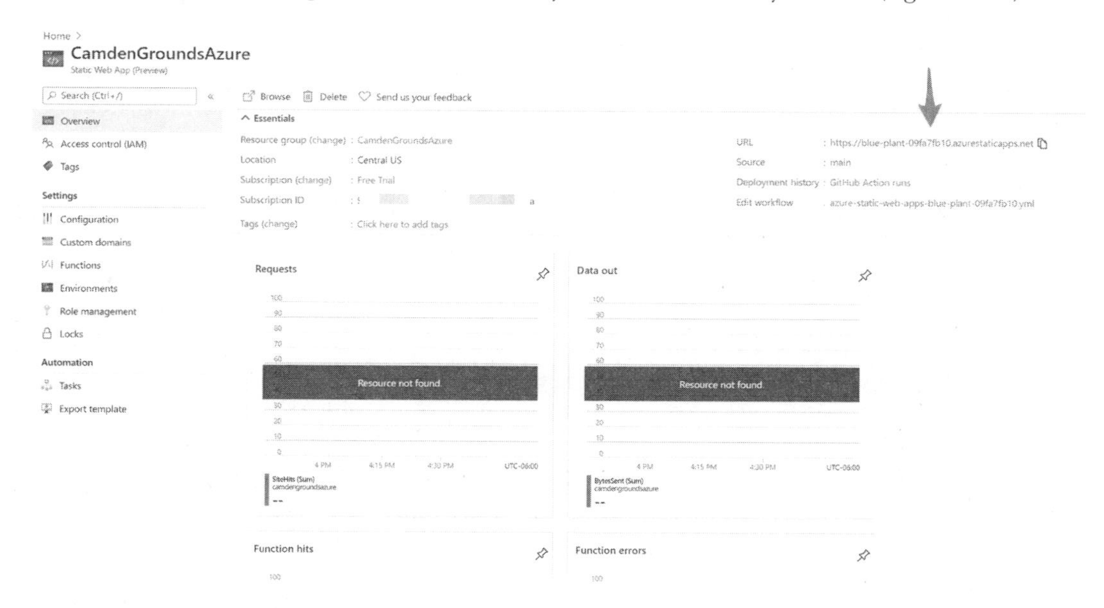

Figure 6.20 Azure portal for the site

If you open that link, you should see Camden Grounds again. Here's the kicker though: if you edit your Eleventy site and commit the change to GitHub, Azure's integrations will kick in, get the latest code, run the build command, and deploy again—all without you lifting a finger!

6.4 Deploying with Vercel

Vercel is the first of two companies we will look at that are focused specifically on serving the Jamstack and web development community in general. Vercel (https://vercel.com/), formerly known as Zeit, offers a variety of services useful for Jamstack developers needing a place to host their sites. A brief list of these services includes the following:

- Deployment from source control providers (GitHub, GitLab, Bitbucket). Like the Azure Static Web App service, this means you can commit code to your repository and Vercel can then automatically update your production site.
- Automatic preview builds for pull releases and branching, making it easy to share and test changes to your site.
- Automatic CDN usage makes your site available globally.
- Serverless function support (you will see an example of this in chapter 8) to add additional functionality to your site.
- A really generous free tier (https://vercel.com/pricing) that makes it easy to see if Vercel is a fit for you.

There are certainly a lot more. But I want to call out a feature of Vercel that I think really sets it apart from others. Vercel has built-in support for over 30 different Jamstack platforms. That support consists of recognizing the framework in use and doing the right thing to build and serve the site. From a practical sense, this means you can go into your project directory, deploy, and Vercel knows what to do. That is incredibly powerful when you need to quickly share a site online for others to look at and test. Vercel still lets you configure your projects manually, which means it will support custom implementations, but the fact that you can go into a Jamstack project, deploy a site, and have it online in a minute or two is awesome.

Let's give this a quick test. Head to the Vercel home page, and click to sign up (figure 6.21). Note that you will be required to use a login on one of the three source control providers Vercel uses.

After you've signed up for Vercel, you can use their web-based dashboard to import projects from source-control providers, but what we want to test is the quick deployment option available via the CLI. The CLI instructions (https://vercel.com/docs/cli) explain in detail all the various options available from the tool, but for now, start off by installing it by using the command:

```
npm install -g vercel
```

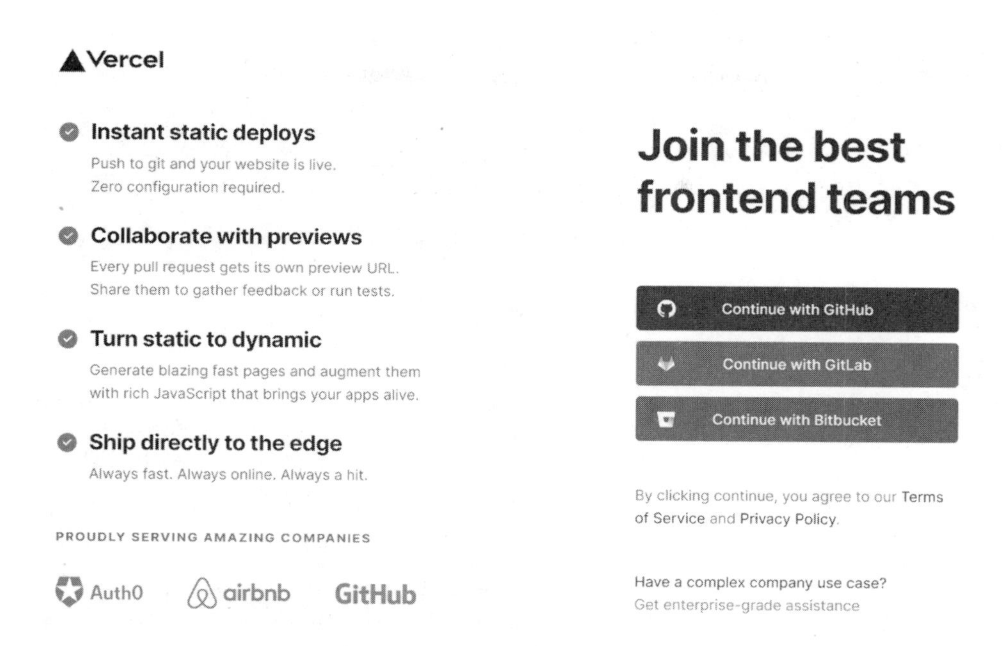

Figure 6.21 Signup options for Vercel require a GitHub, GitLab, or Bitbucket account.

Next, you need to configure the CLI for your account. Do that by running the command:

```
vercel login
```

You'll be asked for the email account associated with your Vercel account. Enter that, and the CLI will fire off a process to send you an email (figure 6.22).

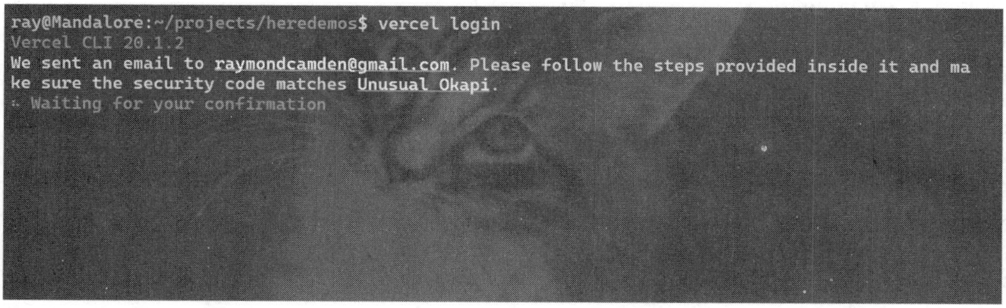

```
ray@Mandalore:~/projects/heredemos$ vercel login
Vercel CLI 20.1.2
We sent an email to raymondcamden@gmail.com. Please follow the steps provided inside it and ma
ke sure the security code matches Unusual Okapi.
: Waiting for your confirmation
```

Figure 6.22 The Vercel login process via the CLI sends you an email to handle confirmation.

As soon as you confirm via the link sent to you in email, the CLI lets you know it's ready to go (figure 6.23).

Log in, and then give Vercel a quick test by using it to deploy the Jekyll site we built in chapter 3. In your terminal, change directories to where you built (or downloaded

```
ray@Mandalore:~/projects/heredemos$ vercel login
Vercel CLI 20.1.2
We sent an email to raymondcamden@gmail.com. Please follow the steps provided inside it and ma
ke sure the security code matches Unusual Okapi.
∴ Waiting for your confirmation
✓ Email confirmed
Congratulations! You are now logged in. In order to deploy something, run `vercel`.
💡 Connect your Git Repositories to deploy every branch push automatically (https://vercel.li
nk/git).
ray@Mandalore:~/projects/heredemos$
```

Figure 6.23 The configured and ready-to-go Vercel CLI

from the book's GitHub repository) the final demo. For me, this is ~/projects/the-jamstack-book/chapter3/startbootstrap-clean-blog-jekyll-master. Once in that directory, deploy the site with `vercel deploy`. While there are multiple options you can use with the CLI, we really want to point out how powerful the defaults are.

You will first be prompted if you wish to set up and deploy the current directory (figure 6.24). Just click Enter.

```
ray@Mandalore:~/.../the-jamstack-book/chapter3/startbootstrap-clean-blog-jekyll-master$ vercel deploy
Vercel CLI 20.1.2
? Set up and deploy "~/projects/the-jamstack-book/chapter3/startbootstrap-clean-blog-jekyll-master"? [Y/
n]
```

Figure 6.24 The deploy process begins by double-checking with you to confirm the current directory is something you want to deploy.

Next it will ask you what scope you are deploying. This is related to your current logged-in account and will differ based on your information. Just click Enter to accept the default.

The next prompt asks if you wish to connect to an existing project. As we are building something new, once again take the default (N) and click Enter (figure 6.25). If you run the deployment again, Vercel will properly recognize it as an existing project.

```
ray@Mandalore:~/.../the-jamstack-book/chapter3/startbootstrap-clean-blog-jekyll-master$ vercel deploy
Vercel CLI 20.1.2
? Set up and deploy "~/projects/the-jamstack-book/chapter3/startbootstrap-clean-blog-jekyll-master"? [Y/
n] y
? Which scope do you want to deploy to? raymondcamden
? Link to existing project? [y/N]
```

Figure 6.25 Continuing the deployment process and specifying that this is a new project

You will then be asked to name this project. The name will default to the directory name, which in our case is a bit long. The name will be used both as a project name and as a default URL. Later you could assign a "real" domain name to your Vercel site. Let's give it a slightly shorter name of verceltest and click Enter.

Now you will be asked in what directory your code can be found. We started the CLI in the right directory, so you can again click Enter to take the default. The CLI will do a bit of inspection on the code base, figure out it's Jekyll, and then ask you if you want to override any settings before it deploys. Once again, click Enter to accept the defaults (figure 6.26).

```
ray@Mandalore:~/.../the-jamstack-book/chapter3/startbootstrap-clean-blog-jekyll-master$ vercel deploy
Vercel CLI 20.1.2
? Set up and deploy "~/projects/the-jamstack-book/chapter3/startbootstrap-clean-blog-jekyll-master"? [Y/
n] y
? Which scope do you want to deploy to? raymondcamden
? Link to existing project? [y/N] n
? What's your project's name? verceltest
? In which directory is your code located? ./
Auto-detected Project Settings (Jekyll):
- Build Command: `npm run build` or `jekyll build`
- Output Directory: `_site` or `destination` from `_config.yml`
- Development Command: bundle exec jekyll serve --watch --port $PORT
? Want to override the settings? [y/N]
```

Figure 6.26 You're given one last chance to modify settings before Vercel deploys.

Now the Vercel CLI will get to work. It pushes up the code from the local directory. It recognizes that you're running Jekyll and does what it needs to support Jekyll on your site. It builds your site, and at the end you're given a report on where your site was deployed (figure 6.27).

```
ray@Mandalore:~/.../the-jamstack-book/chapter3/startbootstrap-clean-blog-jekyll-master$ vercel deploy
Vercel CLI 20.1.2
? Set up and deploy "~/projects/the-jamstack-book/chapter3/startbootstrap-clean-blog-jekyll-master"? [Y/
n] y
? Which scope do you want to deploy to? raymondcamden
? Link to existing project? [y/N] n
? What's your project's name? verceltest
? In which directory is your code located? ./
Auto-detected Project Settings (Jekyll):
- Build Command: `npm run build` or `jekyll build`
- Output Directory: `_site` or `destination` from `_config.yml`
- Development Command: bundle exec jekyll serve --watch --port $PORT
? Want to override the settings? [y/N] n
🔗  Linked to raymondcamden/verceltest (created .vercel)
🔍  Inspect: https://vercel.com/raymondcamden/verceltest/nj79olabj [6s]
✅  Production: https://verceltest-gules.vercel.app [3m]
📝  Deployed to production. Run `vercel --prod` to overwrite later (https://vercel.link/2F).
💡  To change the domain or build command, go to https://vercel.com/raymondcamden/verceltest/settings
ray@Mandalore:~/.../the-jamstack-book/chapter3/startbootstrap-clean-blog-jekyll-master$
```

Figure 6.27 The final output of the deployment to Vercel

The CLI provides a great report of what it did and how long it took. You also get a URL that you can open immediately and see your site. In the test, the URL was https://verceltest-gules.vercel.app but will most likely be different for you.

At this point, you can add a new post, modify an existing post, or make any change you wish. For example, we can modify the title and text of the blog post (_posts/2020-08-24-welcome.html, but your date will differ) to simply add the word *Vercel*:

```
---
layout: post
title: "Welcome to my Vercel blog"
subtitle: "I'm so excited!"
date: 2020-08-24 12:00:00 -0400
background: '/img/posts/01.jpg'
---

<p>
This is my cool Vercel blog!
</p>
```

Once saved, you can run `vercel deploy` again. This time you won't be prompted for anything, but the result is a *preview* build, as you can see in the output in figure 6.28.

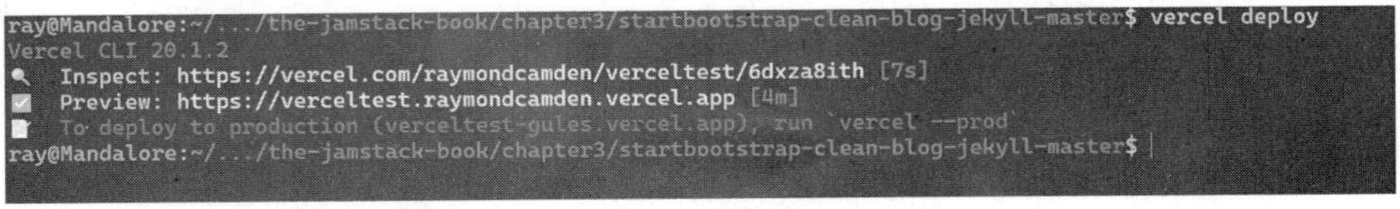

```
ray@Mandalore:~/.../the-jamstack-book/chapter3/startbootstrap-clean-blog-jekyll-master$ vercel deploy
Vercel CLI 20.1.2
🔍  Inspect: https://vercel.com/raymondcamden/verceltest/6dxza8ith [7s]
✅  Preview: https://verceltest.raymondcamden.vercel.app [4m]
📝  To deploy to production (verceltest-gules.vercel.app), run `vercel --prod`
ray@Mandalore:~/.../the-jamstack-book/chapter3/startbootstrap-clean-blog-jekyll-master$
```

Figure 6.28 The result of a preview build

What's really cool about this is that you can now have both URLs open in separate tabs so that you can compare the differences. The CLI also very plainly tells you what to do to have your changes visible in the production build, `vercel --prod`. Running that will be much quicker, as it's moving the preview over to production, and, once done, you can reload the original URL to see your changes (figure 6.29).

Welcome to my Vercel blog
I'm so excited!
Posted by Raymond Camden on August 24, 2020 · 1 min read
VIEW ALL POSTS →

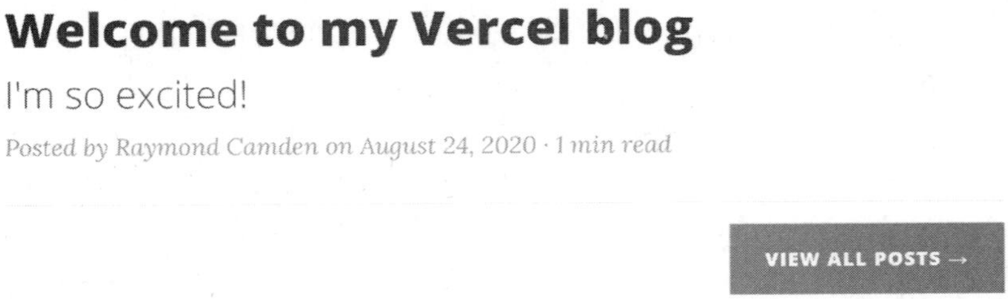

Figure 6.29 Our updated blog post live on Vercel

Before moving on, I want to be clear that this type of updating (changing files, rerunning the CLI) is not something you would use for a production website involving lots of users. In that case you would be using source control. Vercel can listen for changes to your repository and update your sites automatically. But the CLI approach shown here is really useful for testing, sharing sites with others, and so forth. As I said in the beginning, Vercel has a strong set of features, so be sure to peruse its documentation (https://vercel.com/docs) for more information.

6.5 Deploying with Netlify

Let me begin with a bold, if strongly personal, statement. Netlify (https://www.netlify.com/) is the gold standard for hosting Jamstack sites. This is not to say it's perfect or will meet every need. But, for me, it provides the best set of features for the best price and is the standard by which I judge every other service. Let's look at its list of services; as with Vercel, I'm only calling out some of what's available:

- The ability to connect sites to Git repositories and automatically build after a check-in.
- The ability to connect sites to Git repositories and create preview builds based on branches.
- The ability to create complex redirect files. (This is especially useful for people who may move from a traditional application server to the Jamstack and need to ensure older URLs still work correctly.)
- Serverless functions written in JavaScript and Go. You will see more about this in chapter 8.
- Website analytics.
- A custom user identity service that provides user management and security.
- The ability to handle form submissions.
- Special support for large binary files.
- Automatic use of CDNs as well as JavaScript and CSS minification. This also includes the ability to compress images.
- Great support for sites run by multiple people on a team.
- A great CLI for local development.

To be clear, there's more to Netlify than I'll cover here, but you can look at a more comprehensive list on their pricing page (https://www.netlify.com/pricing/).

To test out Netlify, we will once again use the Eleventy demo from chapter 2 and deploy Camden Grounds. If you remember from earlier in this chapter, when we deployed Camden Grounds to Amazon S3, we deployed the output of the Eleventy CLI. If we made a change to the site, we would need to rerun the CLI and copy the files up to S3. While not an onerous task, it's still manual and prone to mistakes. Let's walk through the process of creating a site on Netlify, which will build automatically from a GitHub repository.

If you have not already done so, begin by signing up for Netlify (https://app.netlify.com/signup). They have a free tier that will be more than adequate for your needs. Figure 6.30 shows the multiple ways you can sign up.

After you sign up (and confirm your email), you can then log in. The main interface for Netlify is a dashboard listing all your sites and their current status. As a new user, your dashboard will look pretty empty (figure 6.31).

Figure 6.30 Signing up for Netlify supports multiple login options.

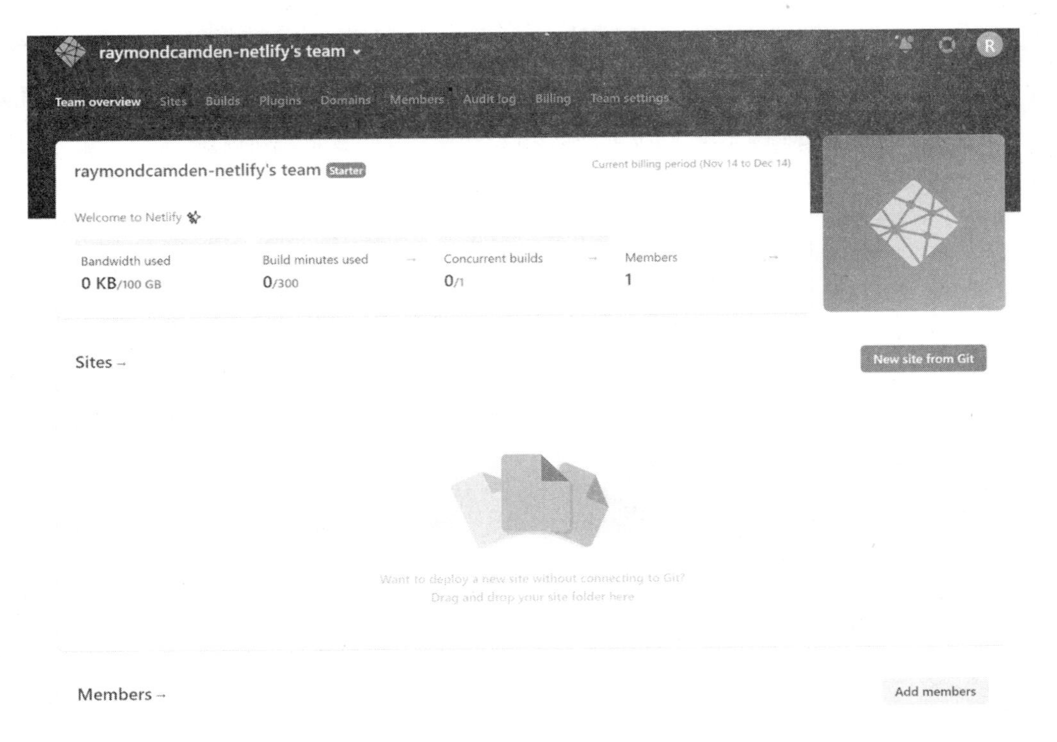

Figure 6.31 The Netlify dashboard for a new user

For comparison's sake, here's my personal dashboard (figure 6.32).

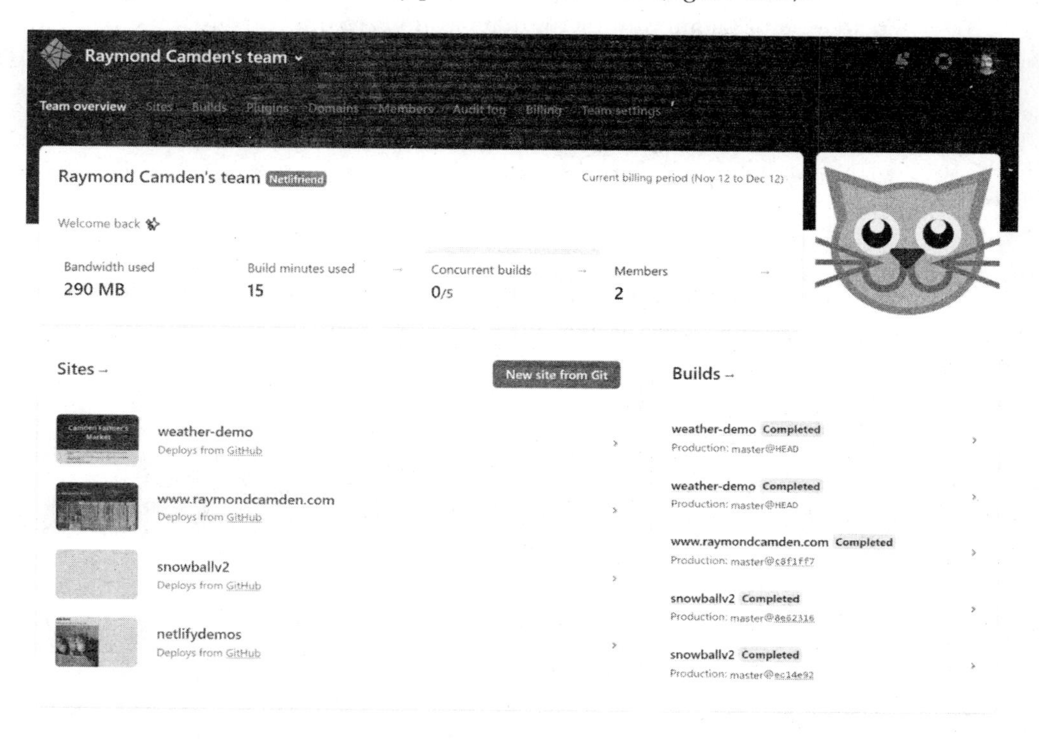

Figure 6.32 A more fully used Netlify dashboard

Notice each site listed contains a screenshot of the current look and feel. On the right is a list of the most recent builds across my entire account. Looking back at figure 6.31, notice the prominent New site from Git button. Let's use this as a way to test Netlify's continuous deployment feature.

While this book has a GitHub repository, in order to test this feature you will need to create your own repository. Most readers are probably familiar with GitHub, but if not, you can read an excellent guide on the GitHub site at https://guides.github.com/activities/hello-world/. Also note that, like Vercel, Netlify has a CLI that supports deploying from your terminal.

As previously mentioned, we're going to use Camden Grounds from chapter 2 as our test. The files for this site are found in the book's repository at chapter2/camdengrounds. In your own GitHub account, create a new repository for Camden Grounds. Once created, simply copy over the files from the previous directory into the new repository. Before you commit these new files, there's one small addition you should make. Git repositories support a way to ignore local directories and files to ensure they don't get pushed to the repository. This is done via a file named .gitignore. When Eleventy outputs a build, it uses a folder named _site by default. Since this is generated output, we do not need it in the GitHub repository. The contents of the file I created before committing Camden Grounds to the repository are provided. Notice we also added `node_modules` to the .gitignore file, shown in listing 6.1. This is required so that Eleventy doesn't try to parse any Node-related files generated during its build.

Listing 6.1 .gitignore file for Camden Grounds

```
_site
node_modules
```

With this in play, commit those files to your new repository. The final result should look like figure 6.33 (although with a different username, of course).

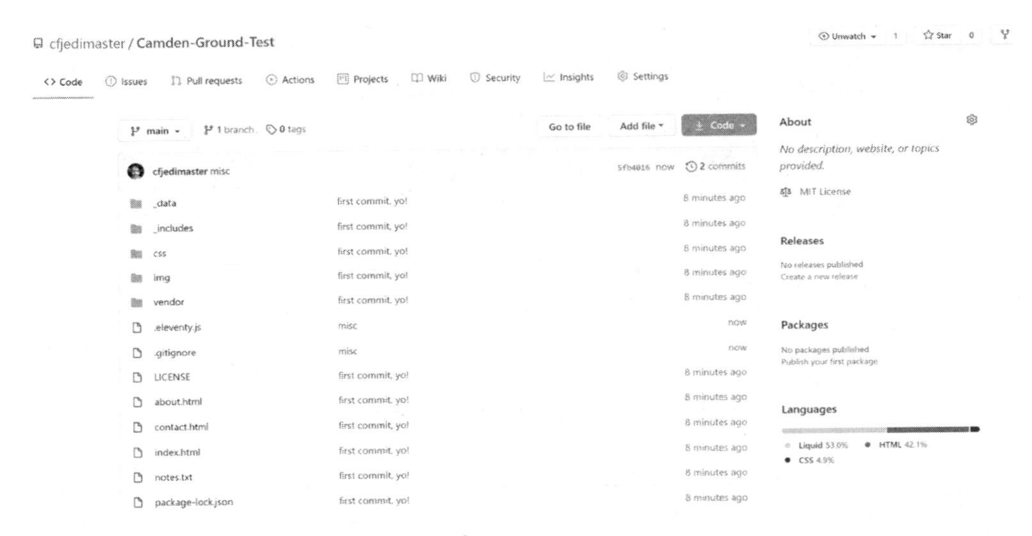

Figure 6.33 The new Camden Grounds repository

If you want to simply copy the repository I used for this test, you can find it at https://github.com/cfjedimaster/Camden-Ground-Test. Now let's hook this up to Netlify!

As mentioned, Netlify prompts you to build a new site from Git, so click that button to begin the process. In figure 6.34, you'll see three different options to begin the process; select GitHub as the provider.

After you've authenticated with GitHub, you will see a list of all your repositories. Find the one you created and click it. The next screen will prompt you for a branch to deploy (defaulting to master or main) and then ask for basic build settings. These build settings are for the command to build your site as well as where the results can be found. To tell Netlify how to build the site, we can use npx, an alternative way of using npm. This is covered in the Eleventy docs (https://www.11ty.dev/docs/usage/) and looks like so: `npx @11ty/eleventy`. By default, Eleventy will output to _site, so we will use that for the publish directory setting. You can see both in figure 6.35.

Create a new site

From zero to hero, three easy steps to get your site on Netlify.

1. Connect to Git provider 2. Pick a repository 3. Build options, and deploy!

Continuous Deployment

Choose the Git provider where your site's source code is hosted. When you push to Git, we run your build tool of choice on our servers and deploy the result.

You can unlock options for self-hosted GitHub/GitLab by upgrading to the Business plan.

Figure 6.34 The beginning of the process to create a new site from a repository

Deploy settings for cfjedimaster/Camden-Ground-Test

Get more control over how Netlify builds and deploys your site with these settings.

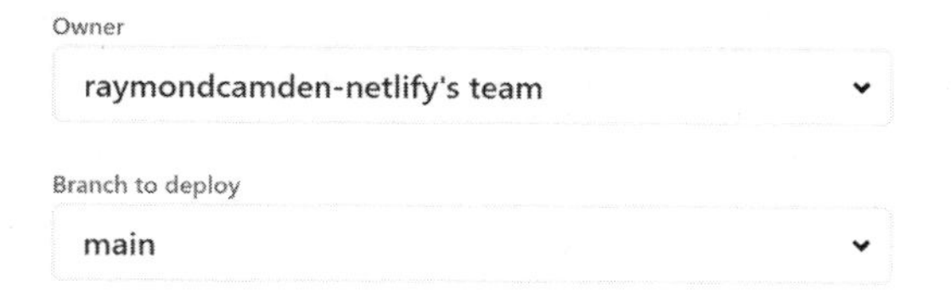

Basic build settings

If you're using a static site generator or build tool, we'll need these settings to build your site.

Learn more in the docs

Build command

npx @11ty/eleventy

Publish directory

_site

Show advanced

Deploy site

Figure 6.35 Configuration settings for the Netlify site

Click Deploy site to begin the process. You can then sit back and watch. The Netlify dashboard will update you about the process of getting your files from GitHub and running your build command. When it's done, you'll see a URL and preview for your site on top of the dashboard (figure 6.36).

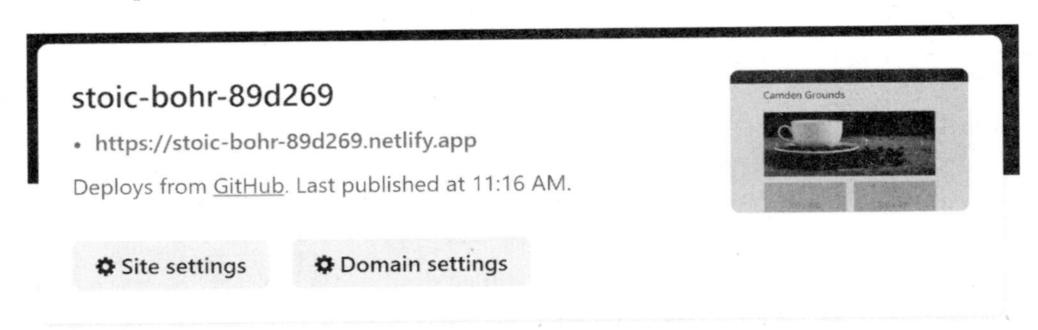

Figure 6.36 The site's updated URL and screenshot

If you click that link, you can browse the site and see Camden Grounds. To truly see the power of the GitHub integration, let's perform a quick test. Open _data/products .json and modify the first product. It doesn't really matter what you change; just make a modification:

```
{
    "name" : "Netlify Powered Coffee!",
    "price" : 2.99,
    "description" : "This coffee is powered by Netlify!",
    "thumbnail" : "http://placehold.it/700x400",
    "image" : "http://placehold.it/900x350"
},
```

Save the file, check it into your GitHub repository, and then watch the Netlify dashboard (figure 6.37).

There are two additional things I want to point out about figure 6.37. Notice the earliest report shows an error. While the nature of the error isn't important, Netlify will handle the error and let you click on the build for details. Also—and this is important—it will *not* break an existing site if something goes wrong in the build. In this particular case it was the first build, so there wasn't a site, but you can rest assured that if you do screw up something, your existing site will continue to work just fine. Netlify only updates the when it's completely, and successfully, built. That means no downtime. Secondly, the message you see under the latest build ("modify product") came right from my GitHub commit. Normally you're supposed to write nice, specific GitHub commit messages, and these will be shown in the Netlify UI.

Once the build is done, you can open the site again and see your new product (figure 6.38).

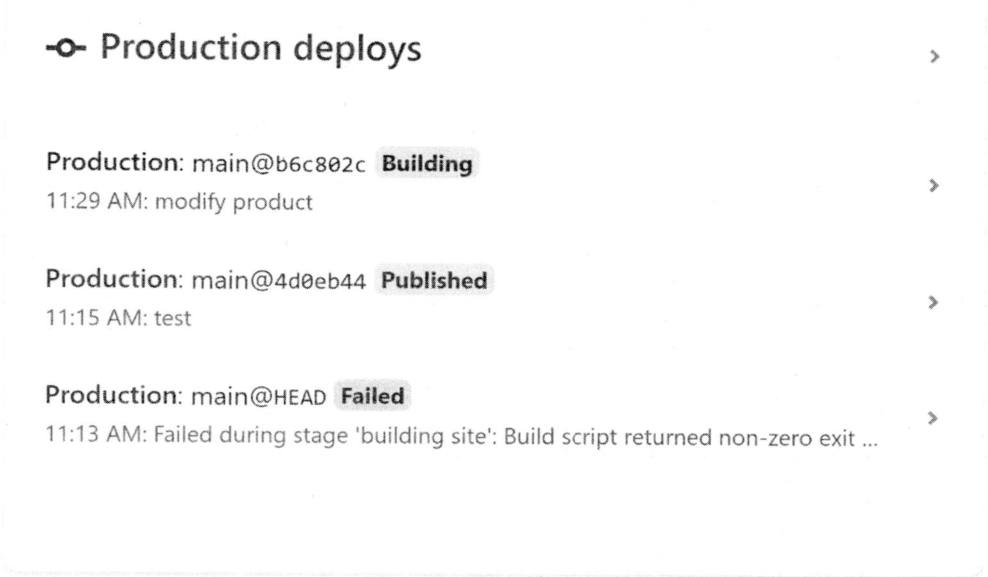

-o- Production deploys

Production: main@b6c802c **Building**
11:29 AM: modify product

Production: main@4d0eb44 **Published**
11:15 AM: test

Production: main@HEAD **Failed**
11:13 AM: Failed during stage 'building site': Build script returned non-zero exit ...

Figure 6.37 The Netlify dashboard showing you information on your build

700 x 400
700 x 400
700 x 400
Netlify Powered Coffee!
$2.99
This coffee is powered by Netlify!
Espresso
$3.99
Lorem ipsum dolor sit amet, consectetur adipisicing elit. Amet numquam aspernatur!
Americano
$5.99
Lorem ipsum dolor sit amet, consectetur adipisicing elit. Amet numquam aspernatur!

Figure 6.38 The updated site based on the last GitHub commit

Let's take a look at one of Netlify's cooler features: form processing. In chapter 7, we'll talk more about handling dynamic aspects of statically generated websites, but since Netlify makes this particular feature so simple to use, we want to cover it here. If you remember, the Contact Us page for Camden Grounds didn't have any real information on it. Let's change that by adding in a form, shown in listing 6.2.

Listing 6.2 The Contact Us form (/contact.html)

```
---
layout: main
title: Contact
---

<div class="row my-4">
  <div class="col">

  <h2>The Contact Page</h2>

  <form>
  <div class="form-group">
    <label for="name">Name</label>
    <input type="text" class="form-control" id="name" name="name">
  </div>
  <div class="form-group">
    <label for="email">Email</label>
    <input type="email" class="form-control" id="email" name="email">
  </div>
  <div class="form-group">
    <label for="comments">Comments</label>
    <textarea class="form-control" id="comments" name="comments">
    </textarea>
  </div>
  <button type="submit" class="btn btn-primary">Submit</button>
  </form>

  </div>
</div>
```

Commit this change to your repository and wait for Netlify to finish building. Once done, you'll see your new form online (figure 6.39). Of course, you can also test how the form looks locally. Typically, you would want to do that so you can quickly see your changes. Since we're building a simple form, we don't need to worry about that right now.

Figure 6.39 A simple contact form

We've got the UI done, but if you actually try to submit the page, nothing happens. Luckily, Netlify makes it pretty simple to handle the form submission. By adding `data-netlify="true"` to your form tag, Netlify will recognize the form submission and automatically save the results. That's all you have to do, but you can configure more as well. Netlify also supports adding captcha and honeypot spam traps to your form. Let's quickly modify contact.html to show this in action. The only change you need to make is to the form tag:

```
<form action="/contact_received.html" data-netlify="true" name="Contact">
```

There are three differences here:

1. First, we specify an action for the form. After Netlify processes and saves the form input, it will then redirect the user to this page. (This page is in the GitHub repository and has a quick thank-you message.)
2. Next, we add the `data-netlify` attribute to tell Netlify to handle the submission.
3. Finally, we provide a name for the form. This will help Netlify recognize different forms if your site has more than one.

Again, commit this change, wait for the build to deploy, and now when you submit the form, you'll end up on the action page. So where did your form data go?

Back in the Netlify dashboard, click the Forms link on top. You'll initially see a list of active forms on the site, which for now is just one (figure 6.40).

Figure 6.40 List of forms that Netlify is handling

Click on the Contact form link to get a list of form submissions. If you click on a submission, you can see the details (figure 6.41).

Netlify offers even more features with forms, including sending a copy of the submission to an email address and passing the data to a serverless function. See the documentation (https://docs.netlify.com/forms/setup/#html-forms) for more information

Figure 6.41 Details of the form submission

about what's possible. In the next chapter, we'll discuss additional ways of handling dynamic aspects of Jamstack sites, including form processing, search, and more.

Summary

- Developers have many options for deploying their Jamstack sites, each with different strengths and weaknesses. Those options include old but still functional vanilla web server installations, cloud-based file hosting providers, and services specifically designed for the Jamstack.
- File-based providers can be simple solutions but do not offer much beyond basic file storage. For basic websites that don't need redirects or other advanced features, a file-based provider may be an adequate solution.
- Microsoft offers a nice option that integrates well with Visual Studio Code. Developers can connect and deploy sites straight from the editor they are (most likely) already using to develop their site.
- Both Vercel and Netlify tailor specifically to Jamstack and frontend developers, providing numerous features that make hosting attractive. Form processing, redirect rules, and serverless functions are just a few examples.

7 Adding dynamic elements

This chapter covers

- Adding dynamic elements back into static web pages
- Processing form submissions via multiple providers
- Creating a search interface to a static site

In the old days (you know, two or three years ago), what we know now as the Jamstack was a bit simpler. Typically we referred to static sites and static site generators. The problem with these terms is that they implied a static, unchanging site that couldn't respond to users' needs. That wasn't the case then and isn't the case now.

There are numerous options (some free, some commercial) that aim to provide interactivity to web pages. In this chapter, you will see different examples of these services as well as how they can be integrated in some of the previous demos. We'll discuss tradeoffs, prices, and other considerations that developers need to be aware of before selecting a particular product.

7.1 Forms, forms, and more forms

One of the first tasks I did as a professional web developer in the ancient days of the 1990s was form processing. Back then, our sites were much like now, simple HTML, and in order to add functionality to a site, like processing a form, we would use programs written in Perl. I had a knack for Perl, so in many projects, I'd focus on that area. Forms have been around since the beginning of the web, and so has the need to process those forms. Let's look at a few different options for adding processing to your Jamstack site. Broadly, we're going to look at two different ways of doing this: with forms hosted elsewhere and embedded on your site, and as services that simply handle receiving the data of a form.

7.1.1 Using Google Forms

Google Forms (http://forms.google.com/) is a free offering from Google related to the more generic Google Docs service. Developers can create forms with multiple types of questions and different styles. Form data is automatically stored in a Google Sheet (their version of Microsoft Excel) and can also be emailed directly to whomever needs to get the results.

Using Google Forms in the Jamstack means creating and designing your form on Google's site, and then adding the embed code to an HTML file. This can be done in non-HTML files as well. For example, if you are using Liquid in your SSG, you can add the embed code there, which will be a part of the final HTML file output when building your site. Let's walk through creating a basic form that will mimic the typical Contact Us form seen on websites.

Begin by going to the Google Forms site (you will need a Google account to use this service). You should see a list of templates and any recent forms (figure 7.1).

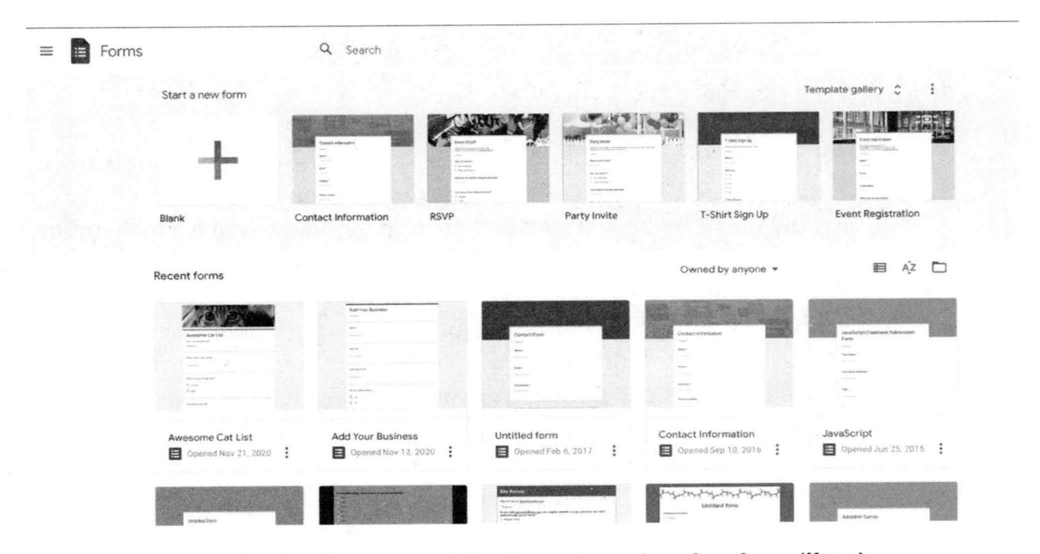

Figure 7.1 The Google Forms home page, listing templates and previous forms (if any)

Start by clicking the Blank template for now. The included templates are pretty nice, but it's best to start off simple. This will take you to the Google Forms editing experience (figure 7.2).

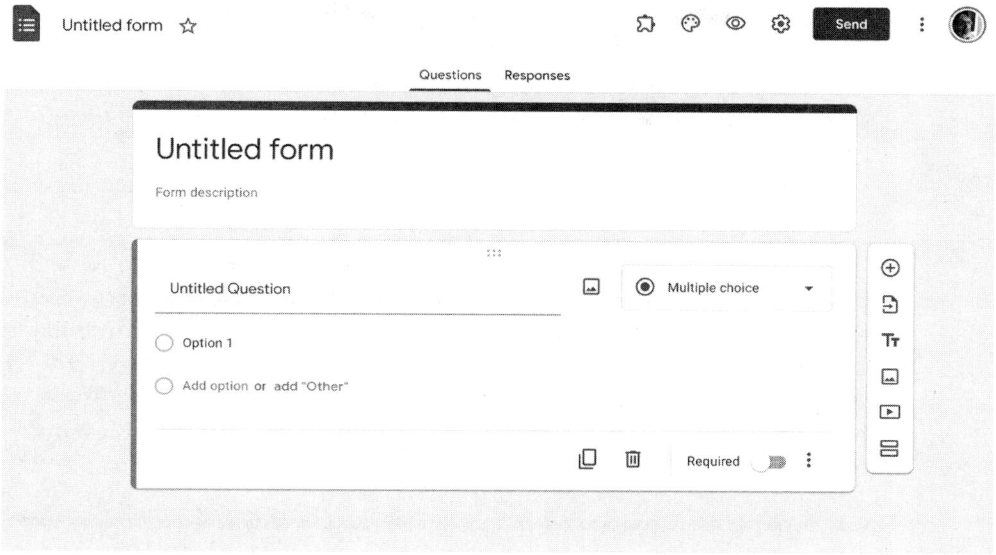

Figure 7.2 The initial blank form

Google's form editor is incredibly well done. You can choose between different types of questions (short answer, longer form, multiple choice, etc.) and are free to enter the question-and-answer text as you see fit. Google does some pretty amazing parsing of your questions as well, enabling you to write answers quicker. For example, if your question implies a yes or no answer, Google picks up on that and suggests them for answers.

As we said, our first example is going to be a contact form. These have been around forever and typically follow a format of asking for your contact information and providing a place to ask your question or send your feedback.

The design of forms can be a complex topic. How many questions do you ask? What questions are required? What language do you use for a particular question? Get any of these wrong and your users may simply go away, or even worse, spend time writing answers and give up halfway. For this particular example, let's use the following questions:

- *What is your name?* This will be a short text answer and will be required.
- *What is your email address?* This will also be a short text answer and will be required, and it is important so that we have a way to reach the user and respond to their comments.

- *Do you like our site?* This will not be required and will be used as a quick way to gauge if users are enjoying using our site. It's optional, so users can skip it, but we should be prepared for results to skew negative. Why? If a user is happy using our site and has no problems, they probably won't bother contacting us to tell us this, so it's fair to assume that a good portion of users who take the time to fill out this form have a problem of some sort and are probably not happy. Again, this is where the complexity of forms comes into play: you not only have to think of the user experience of the person filling out the form but also the psychology of why they are doing so.
- *Your comments.* The final field will be a long text field that is required and will be used by the site visitor to either ask a question or provide feedback.

To begin, first set a name for the form, replacing "Untitled form" with "Contact Form." Optionally, you can enter a form description, but don't worry about that now. Then enter the text of the first question in the first field of the form builder. Be sure to set it to Required. You'll notice that as soon as you enter the text of the question, that intelligence will fire, and Google will default the type to short answer. This is just Google trying to help, but and you can change this to whatever you want. Figure 7.3 shows how this should look when done.

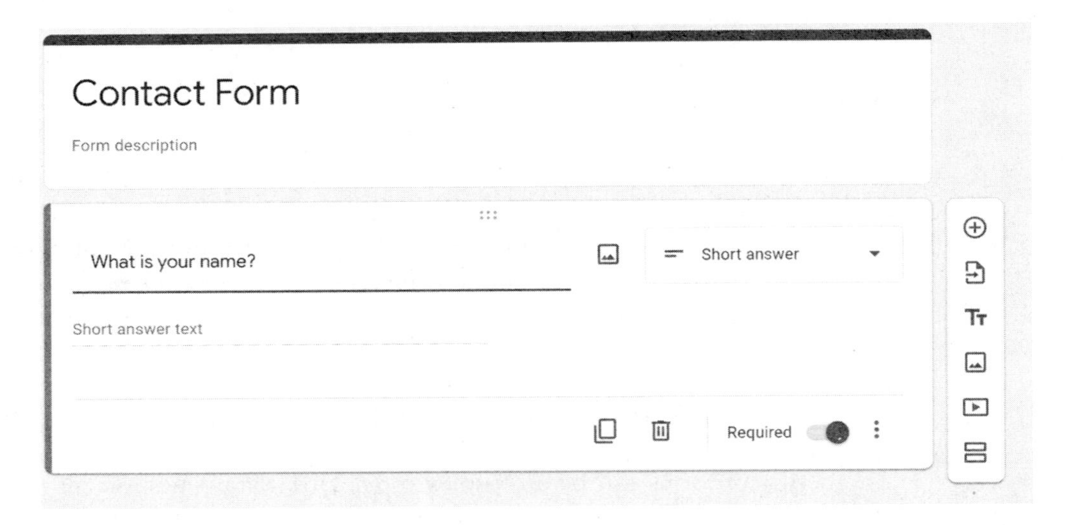

Figure 7.3 The form after a bit of editing

Repeat this step to ask for the user's email address. To add a new question, click the + icon on the right-hand side, as shown in figure 7.3. Then add the question about whether the user likes the site. Not only will Google determine that this is a yes/no question, it will suggest those answers along with "maybe." You can click the suggestions to quickly add them (figure 7.4).

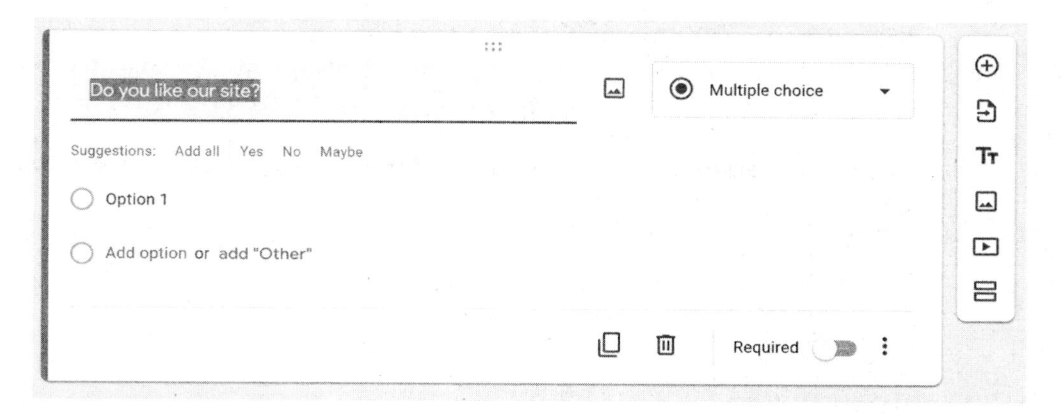

Figure 7.4 Google's intelligent form editor recognizing your question and suggesting answers

After adding yes and no as answers, add the final question. Google should suggest "Paragraph" as the type, but be sure to set it to Required. When done, click the Preview icon on top (it's an eyeball), and a new tab will open with your form (figure 7.5).

Contact Form

* Required

What is your name? *

Your answer

What is your email address? *

Your answer

Do you like our site?

Yes

No

Your comments. *

Your answer

Submit

Never submit passwords through Google Forms.

This content is neither created nor endorsed by Google. Report Abuse - Terms of Service - Privacy Policy

Google Forms

Figure 7.5 The complete form

This preview is a fully functioning form that you can use right now and submit. You can see how the validation logic works by intentionally leaving things out. Go ahead and submit your form. When you do, you'll get a simple confirmation (figure 7.6).

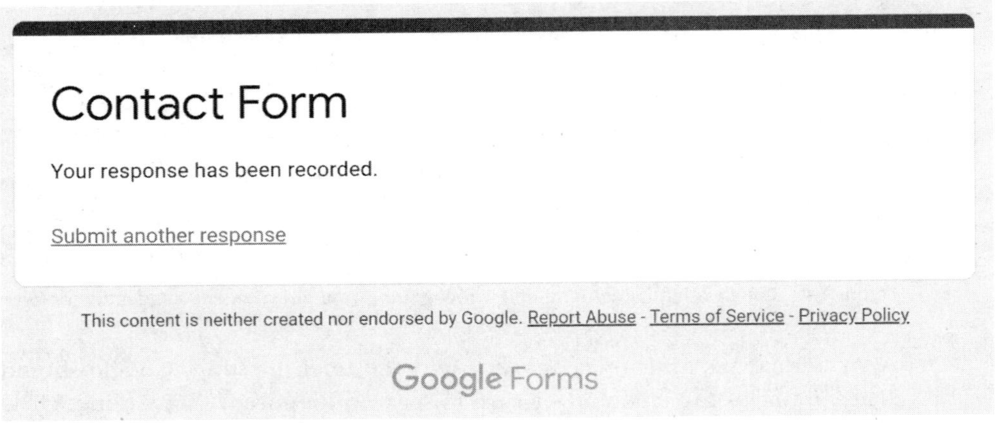

Figure 7.6 What users see after submitting the form

Do this a few times and then return to the tab where you're editing the form. You'll see the Responses tab will notice that submissions have been received (figure 7.7).

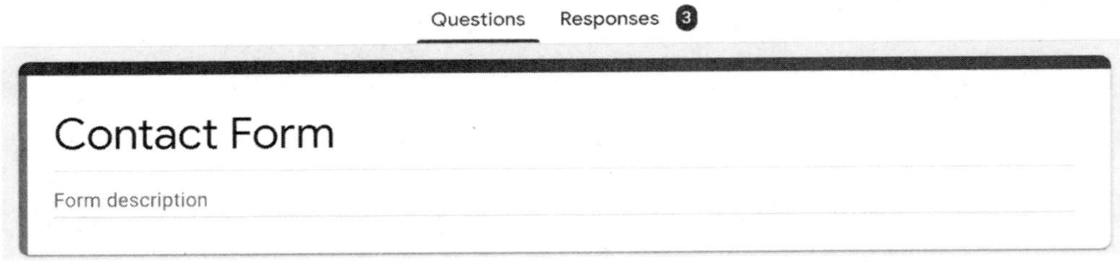

Figure 7.7 Highlighting the Responses tab

Clicking on the Responses tab will show you a summary of the responses (figure 7.8). Google does an admirable job of displaying this data and recognizes that the "Do you like our site?" question makes sense as a pie chart, but oddly displays email addresses as a bar chart.

Clicking on the Question tab will let you look at results for one question at a time, whereas Individual shows you one complete response at a time. The green icon on the top lets you create a Google Sheet and automatically connects the response (and future ones too!) to a spreadsheet. In most cases, though, a developer (or the owner of the website) will prefer to get an email response. To set this up, click the three-dot

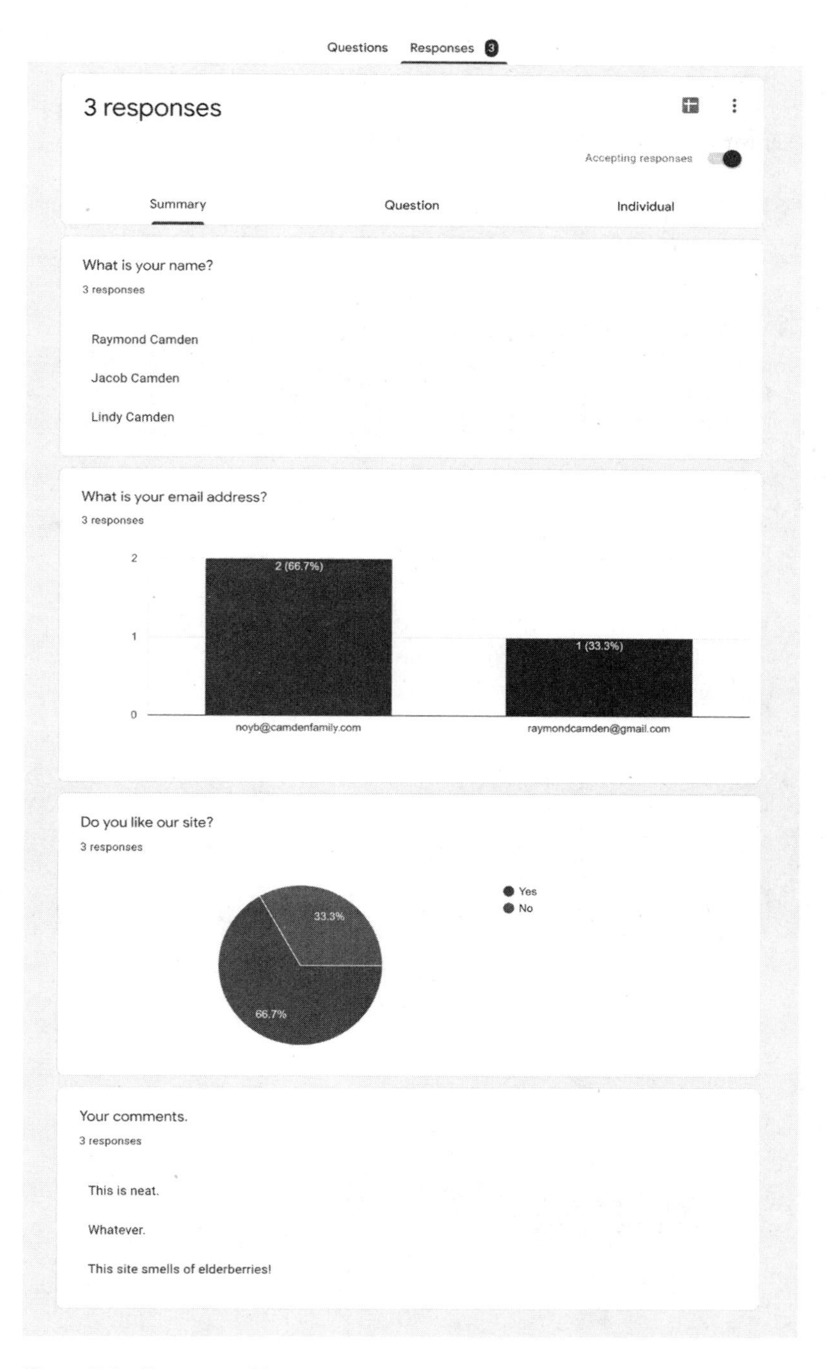

Figure 7.8 Summary of form responses

menu to the right of the green spreadsheet icon, and in the pop-up menu, select Get email notifications for new email responses (figure 7.9).

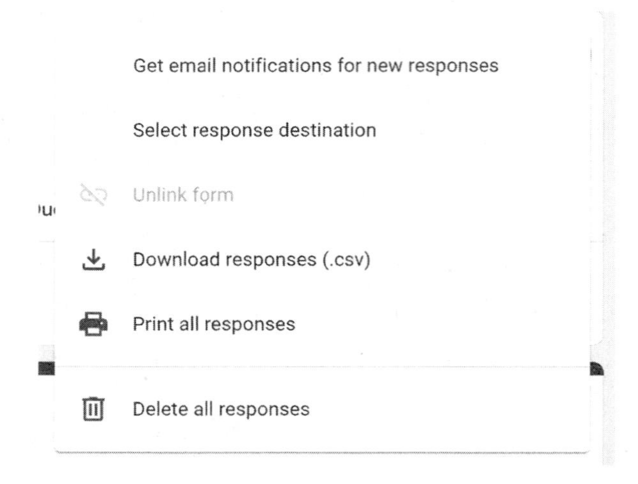

Figure 7.9 Setup and configuring email responses

Clicking this will enable email notifications for future submissions, but note that it will go to the owner of the Google account. Now, if you fill out the form again, you'll receive an email (figure 7.10).

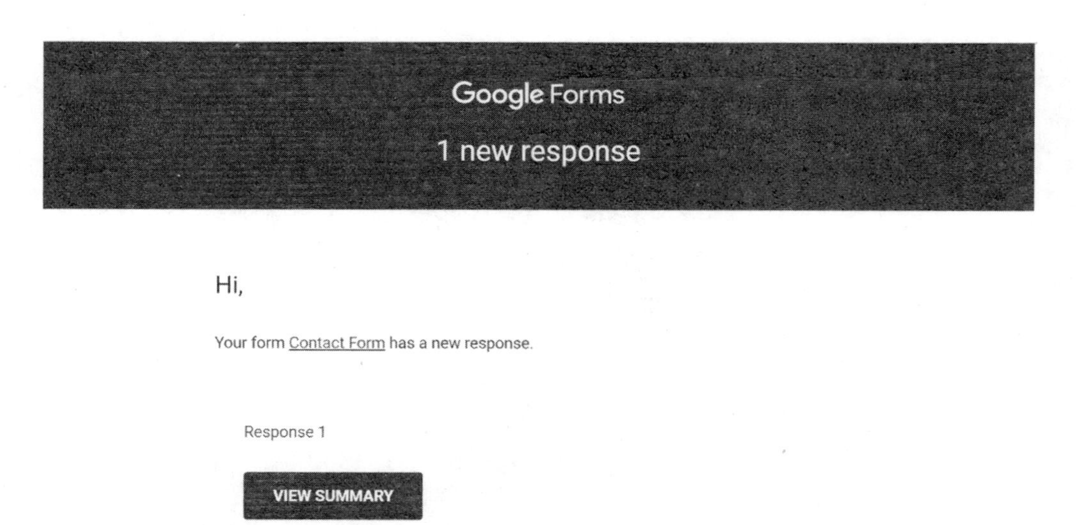

Figure 7.10 Email received after the form has been submitted

Unfortunately you still have to click to get to the responses, but at least you know a response was submitted.

Now that your form is built and it can take in submissions, what's the next step? While our website could simply link to the form, that would mean the user is leaving our site, and generally that's not a desirable thing. Instead, we can put the form directly on our site. This can be done via an iframe embed code. It's not terribly obvious where you get that code, but it's found via the Send button on the top of the form editor. Clicking this opens a dialog that defaults to sending the form via email, but clicking the "<>" icon will show you the iframe code (figure 7.11).

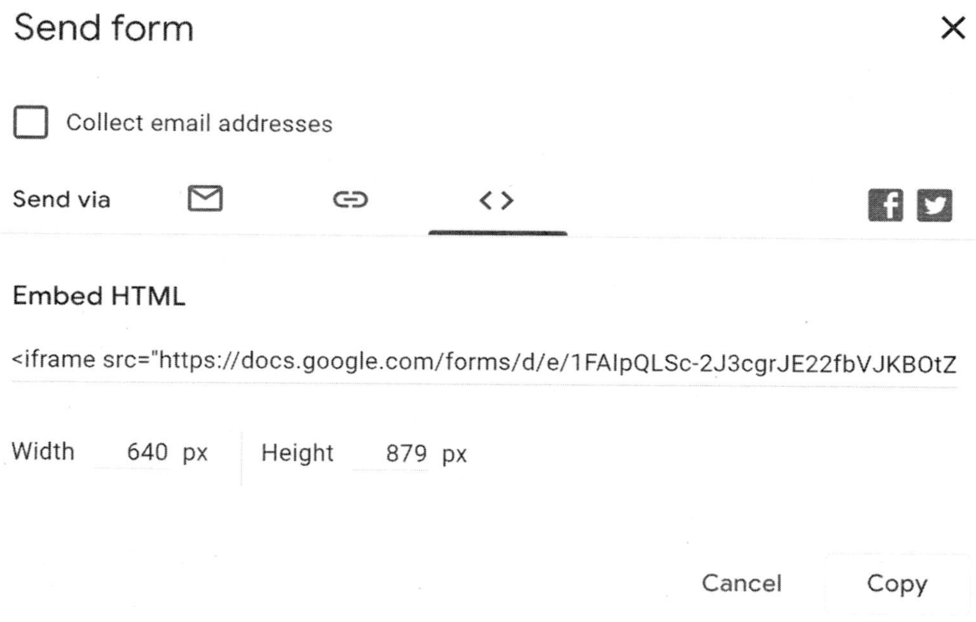

Send form

Collect email addresses

Send via

Embed HTML

<iframe src="https://docs.google.com/forms/d/e/1FAIpQLSc-2J3cgrJE22fbVJKB0tZ

Width 640 px Height 879 px

Cancel Copy

Figure 7.11 The Send dialog's HTML option

With this code, you can now embed it in any web page you want, including within a Jamstack site. But let's begin with a simple example using a typical HTML page with layout. The embed code from Google Forms is included as part of the page. Be sure to replace this with your own embed code.

Listing 7.1 A Google Form embed example (/chapter7/forms/test1.html)

```
<!DOCTYPE html>
<html>
<head>
<title>A Regular Page</title>
<style>
body {
```

```
    background-color: bisque;
    font-family: Verdana, Geneva, Tahoma, sans-serif;
}

footer {
    margin-top: 10px;
    font-size: 10px;
}
</style>
</head>

<body>

    <h1>My Site's Contact Form</h1>

    <iframe src="https://docs.google.com/forms/d/e/1FAIpQLSc-
    ➥ 2J3cgrJE22fbVJKBOtZFlmzpiqE36SE-eAmJHqTh6NZHrA/viewform?embedded=true"
    ➥ width="640" height="879" frameborder="0" marginheight="0"
    ➥ marginwidth="0">Loading…</iframe>

    <footer>
    <p>Copyright Whenever</p>
    </footer>

</body>
</html>
```

This is a rather short template with a header and footer and the iframe from Google Forms inside. We also use a bit of CSS here to style the page. The only reason for this is to show how the form looks when added to a page that has its own unique style. Google Forms allows for some customization in look and feel, so you can expect to have to work with that a bit, but we wanted to demonstrate an out-of-the-box integration with no customization. Figure 7.12 shows how this renders on a page.

As you can see, it integrated just fine. There is a scrollbar, but that could be adjusted by modifying the iframe's height attribute or using CSS. If you submit this form, the process is done entirely within the iframe, and the user never leaves the site. While it definitely sticks out a bit on the page, this is a quick and simple solution. Even better, the form could be edited by a nontechnical user, and since it's an embed, nothing would need to change on your site to reflect their changes. Now that you've seen an example of a remote, hosted form, let's consider something a bit more integrated.

My Site's Contact Form

Contact Form

* Required

What is your name? *

Your answer

What is your email address? *

Your answer

Do you like our site?

Yes

No

Your comments. *

Your answer

Submit

Never submit passwords through Google Forms.

Google Forms This content is neither created nor endorsed by Google.

Copyright Whemever

Figure 7.12 The Google Form

7.1.2 Integrating FormCake

While Google Forms works via an external host displayed in an iframe, developers may want more control over the design and setup of their forms. Multiple services now exist that provide an endpoint, or a place to send your form data to, that will then take in the data, do "stuff" (what they do depends on the service), and then redirect the user back to your site. For most people, they have no idea what's happening. They simply clicked your Submit button and then were presented with a thank you or confirmation page. But behind the scenes, the service in question parsed the form, did something with that data, and then redirected the website visitor right back to your site.

One of these services is FormCake (https://formcake.com). This service provides form processing and includes things like file upload support, spam protection, and the ability to perform actions on the data. At a minimum, it can email the data to you (or the owner of the site), but it can also integrate with solutions like Zapier (an automation service that lets you connect different apps in workflows, for example, on a form submission sending information to Salesforce). FormCake currently has three different tiers of pricing (see more at https://formcake.com/pricing), but their free tier allows for unlimited forms, 100 submissions, and basic spam protection. This is enough for us to test, so let's give it a shot.

Begin by signing up (you can use either an email and password combination or your GitHub account). After you've done that, you are taken to a dashboard with one form already created (figure 7.13).

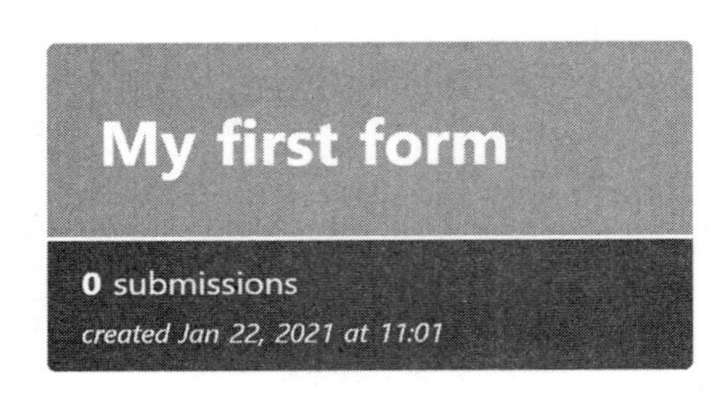

Figure 7.13 The forms dashboard at FormCake

Clicking into the form gives you many settings and integration instructions for working with it (figure 7.14).

Figure 7.14 Instructions on how to use the form

The instructions mostly boil down to ensuring your form uses `POST` and copying the action endpoint shown in the first step. Let's build our form that uses this action. The iframe from the Google form has been replaced by a form entirely written by hand and using the endpoint provided by FormCake. It asks the same questions as before.

NOTE The endpoint is specific to the author and should be replaced with the one you get from FormCake.

Listing 7.2 The FormCake form (/chapter7/forms/test2.html)

```
<!DOCTYPE html>
<html>
<head>
<title>A Regular Page</title>
<style>
body {
    background-color: bisque;
    font-family: Verdana, Geneva, Tahoma, sans-serif;
}

footer {
    margin-top: 10px;
    font-size: 10px;
}
</style>
</head>

<body>

    <h1>My Site's Contact Form</h1>

    <form method="post" action =
    "https://api.formcake.com/api/form/451d4b55-3e2a-4eee-b7c4-
    ➥ 1b54041a5365/submission">

    <p>
    <label for="name">What is your name?</label>
    <input name="name" id="name" required>
    </p>

    <p>
    <label for="email">What is your email address?</label>
    <input name="email" id="email" type="email" required>
    </p>

    <p>
    Do you like our site?<br/>
    <input name="like" id="likeyes" value="yes" type="radio">
    <label for="likeyes">Yes</label><br/>
    <input name="like" id="likeno" value="no" type="radio">
    <label for="likeno">No</label><br/>
    </p>
```

```
    <p>
    <label for="comments">Your comments</label>
    <textarea name="comments" id="comments" required></textarea>
    </p>

    <p>
    <input type="submit">
    </p>

    </form>

    <footer>
    <p>Copyright Whenever</p>
    </footer>

</body>
</html>
```

If you fire up a local web server to test this (I suggest https://www.npmjs.com/package/httpster), you can view the form, submit it, and then end up on a default submission page for FormCake (figure 7.15).

Thank you!
Your submission has been recieved.
Powered by formcake
Learn More

Figure 7.15 The default FormCake response

If you go back to the FormCake dashboard, you can click on the Submissions tab and see your response (you may need to refresh). Clicking on it will give you a detailed view (figure 7.16).

To make this form a bit nicer, let's make two changes in the form settings. In the Settings tab, scroll down to "Success Redirect" and add a URL to redirect the user back to where they submitted the form. Right now we're testing locally, so we can use a localhost URL, but that would not work in production. In my environment, my form was available at http://localhost:3333/chapter7/forms/test2.html. I created a new file

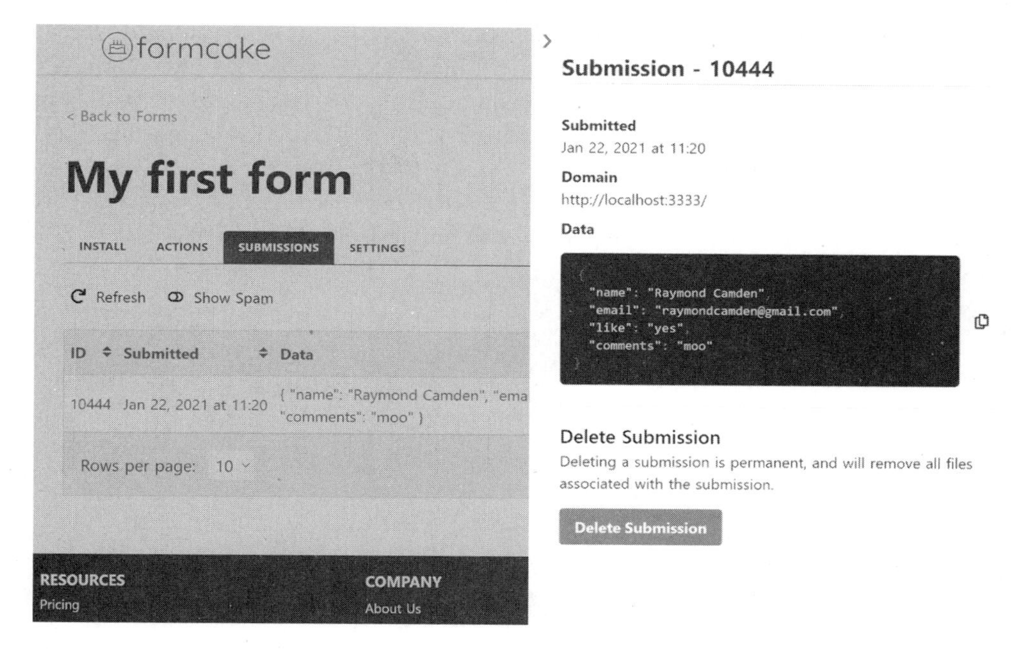

Figure 7.16 FormCake's submission view

named test2_thankyou.html (you can find this in the GitHub repository for the site. It isn't really important what's in the file; it just needs to exist), and then used this as the redirect value: http://localhost:3333/chapter7/forms/test2_thankyou.html (figure 7.17).

Edit Form

Name *

My first form

Description

Description

Honeypot Field

Submission data with a value for this field are marked as spam.

Honeypot Field

Success Redirect

Redirects are used when posting to your form endpoint using a <form> tag's "action" attribute

http://localhost:3333/chapter7/forms/test2_thankyou.html

Figure 7.17 Form settings with the redirect specified

Next, we need to set it up so you can get an email every time the form is submitted. Click on the Actions tab, and in the Add an Action drop-down menu, select Email Notification

Action. This will prompt you for a name for the action, an email address to send the information to, and a subject. Enter "Email Notification" for name, your own email address, and "Form Submission" for the subject. Now you can fill out your form again, and after submitting, you will be redirected to your thank-you page. The end user will not see a FormCake site and won't even know you're using the service. To be clear, it isn't hidden, and developers could easily monitor the network traffic, but casual users won't know (or care). Soon after, you'll get an email notification (figure 7.18).

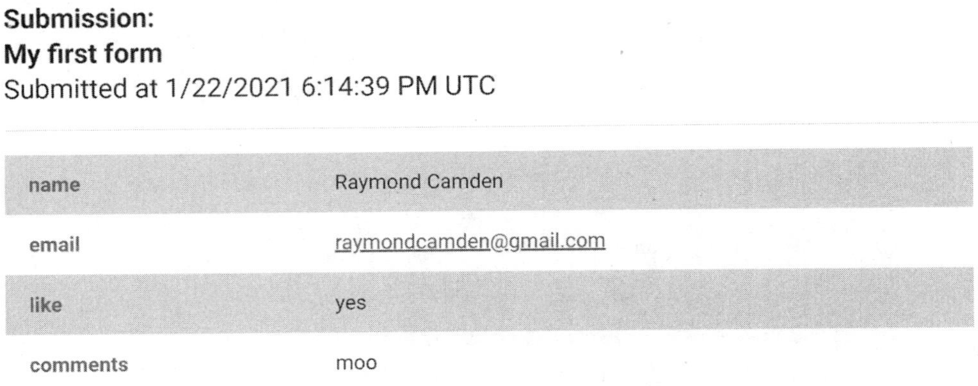

Submission:
My first form
Submitted at 1/22/2021 6:14:39 PM UTC

name	Raymond Camden
email	raymondcamden@gmail.com
like	yes
comments	moo

Figure 7.18 Email copy of the form submission

If you want something a bit more custom, FormCake even allows you to define an email template that uses Liquid for variable replacement. This is a pretty powerful feature!

7.1.3 Other options

There are, of course, other options for working with forms. There are numerous ones similar to FormCake, where you simply use a specific action for your form. There's also Wufoo (https://www.wufoo.com/), which acts more like Google Forms but has powerful design and editing features. As mentioned in the previous chapter, Netlify as a Jamstack host has built-in form processing. In the next chapter, we'll discuss serverless functions, and that's yet another way you could respond to a form post.

7.2 Adding search

After processing basic form input, search is probably one of the most important features for a Jamstack site. If your site has more than a few pages, giving users the ability to quickly find what they want becomes important. In this section, we'll discuss two different options for search and will follow a format much like the previous section. We'll begin with a "drop into place" solution, once again from Google: Programmable Search Engine (https://programmablesearchengine.google.com/about/).

Google's Programmable Search Engine was previously known as Custom Search Engine, so if you follow up the tutorial in this book with more research, you may find articles referring to that, or CSE. This is a service from Google that lets you define, essentially, a portion of Google's search engine to use within your own site.

To begin, simply click the Get Started button from the Programmable Search Engine site and log in with your Google credentials. Once logged in, you'll be taken to the dashboard (figure 7.19).

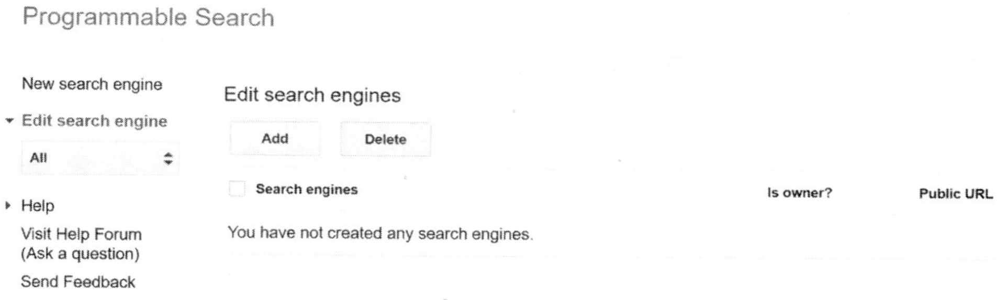

Figure 7.19 The Programmable Search dashboard

Begin working with this service by clicking the Add button. The first prompt you'll be presented with is the sites to search, and here's where things get interesting. You can enter any site you want here. That's right—even if you are building your own site at a domain X, you can enter domain Y (and more) for your search engine. You'll probably want to use your own domain, but Google lets you decide whatever makes sense. Note (as seen in figure 7.20) that you can enter subdirectories as well.

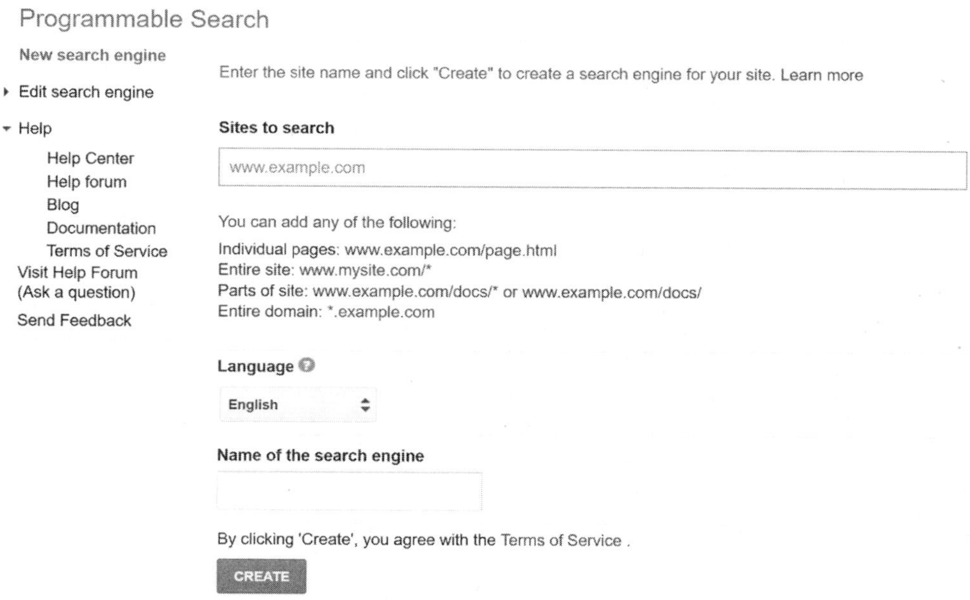

Figure 7.20 The initial screen for setting up a programmable search engine

As we said, you can enter anything you'd like here. If you've already deployed your own Jamstack site, even to a temporary location, you can enter the URL. For the purposes of this book, and to give us a lot of content to use, we'll use my blog at raymondcamden.com. For the name of the search engine, let's use JamstackSearch1. That's purely arbitrary and is only used to help differentiate one of your programmable search engines from another. In a real-world scenario, I would typically use the same name as the site itself. After you've entered your values, click the Create button. You'll then see a success message (figure 7.21).

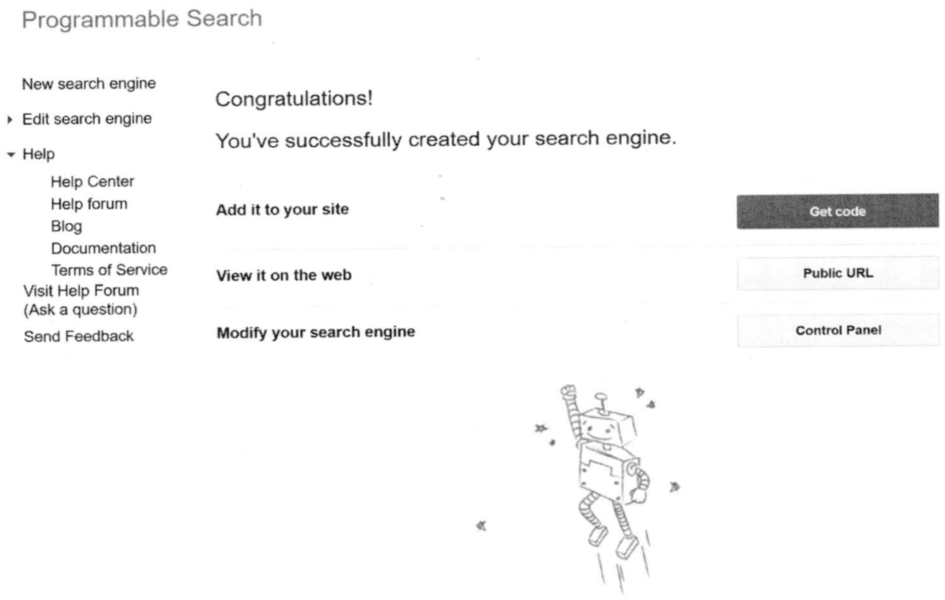

Figure 7.21 Congratulations, you've built your own search engine (well, Google did, to be honest).

At this point, note the three options. The first (Get code) will give you the code needed to add the search to your site, and we'll get to that in a second. The third option (Control Panel) is where you'll configure options; we'll demonstrate that too. The one in the middle (Public URL), though, is something else. As soon as you build a programmable search engine, Google provides a URL that lets you use your search engine right away. This is great if you want to test how well it's working, how it ranks results, and so forth. You could also share this with your client while you are building your site so they can see it as well.

For now, click on Get code. In listing 7.3, you'll be presented with a short code snippet. You'll take this code and drop it into a simple HTML page. The same basic shell from the previous section was used.

Listing 7.3 A search test page (/chapter7/search/test1.html)

```
<!DOCTYPE html>
<html>
<head>
<title>A Regular Page</title>
<style>
body {
    background-color: bisque;
    font-family: Verdana, Geneva, Tahoma, sans-serif;
}

footer {
    margin-top: 10px;
    font-size: 10px;
}
</style>
</head>

<body>

    <h1>Search</h1>

    <script
    async src="https://cse.google.com/cse.js?cx=9fda4c9699117d517"></script>
    <div class="gcse-search"></div>

    <footer>
    <p>Copyright Whenever</p>
    </footer>

</body>
</html>
```

The beginning of the snippet from Google

The end of the snippet

As with Google Forms, you can drop in the snippet where it makes sense to you. Fire up a local web server and then run the page. You'll see a basic form experience completely powered by Google (figure 7.22).

Figure 7.22 The default search experience provided by Google

Enter something in the search. Depending on what domain you entered, you should try something that makes sense. If you used my domain (raymondcamden.com), you can try "vue.js" as your input. Figure 7.23 demonstrates what you may see.

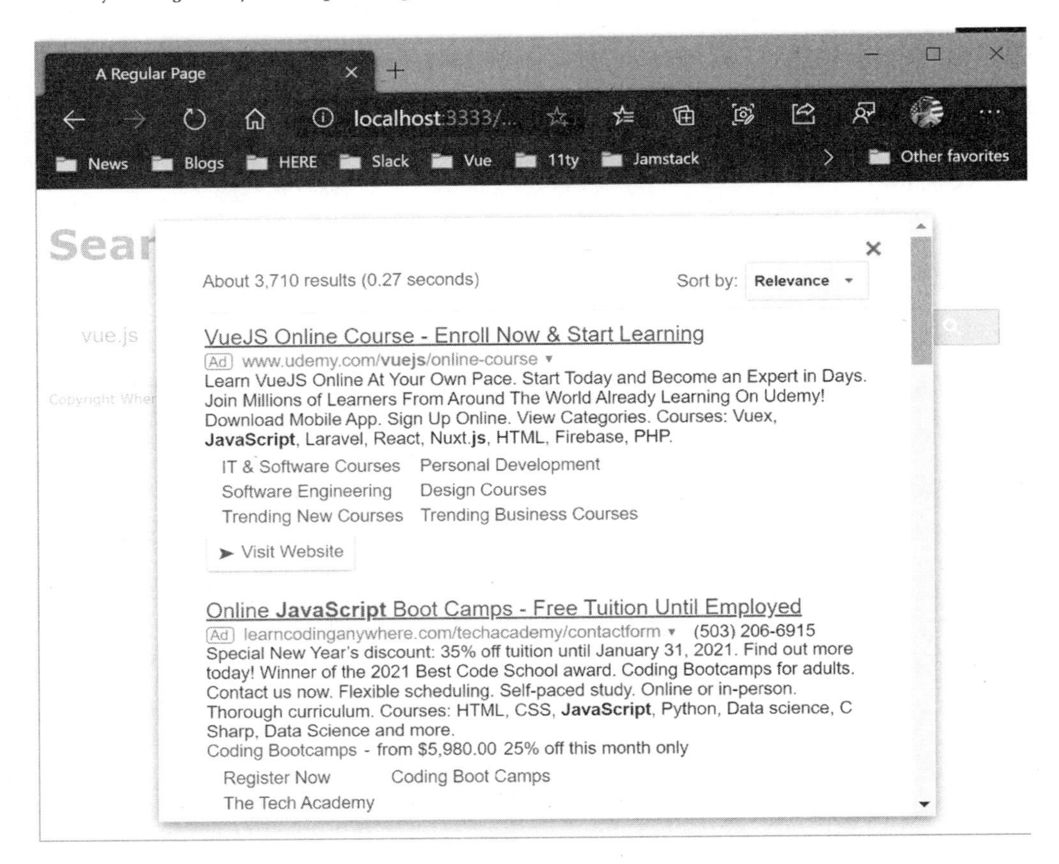

Figure 7.23 Results from searching the programmable search engine

Right away you'll notice that the UI is a bit different from what you may expect. Instead of displaying the results on the page, a floating modal is used. If you don't like that (I certainly do not), we can tweak it, and we'll do so in a moment. The next thing you may notice, especially in figure 7.23, is that *every visible result* is an advertisement. Yep, just like the main Google.com site, you'll get ads in your result. Not shown in the figure is that if you scroll past the four ads (yes, four, and your results may vary, of course), you do get good results with snippets, and images at times (figure 7.24).

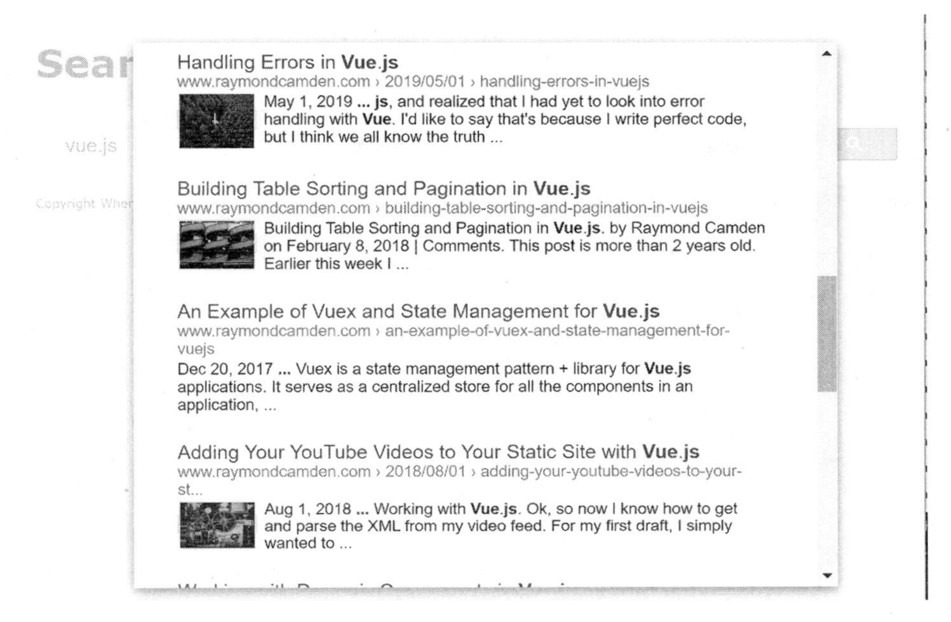

Figure 7.24 "Real" results can be found after the ads. Google does need the money, you know.

Let's start customizing the search engine. Back in the dashboard, click on Look and feel. Here you will find multiple options for customizing how the search engine is displayed, but let's start by changing the layout from "Overlay" to "Full width" (figure 7.25).

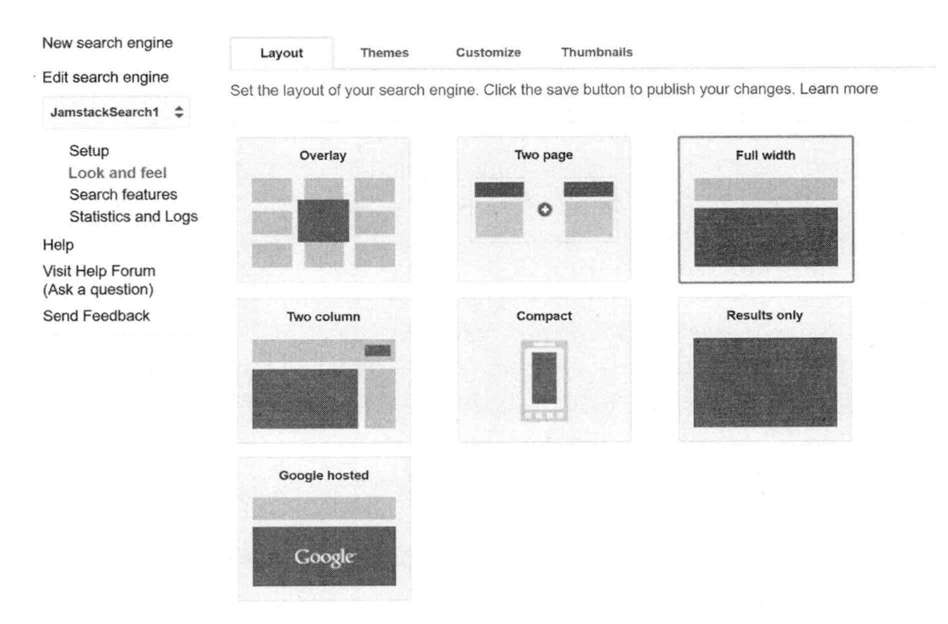

Figure 7.25 Specifying a new layout for your search engine

Not shown in figure 7.25, but handy to use, is a live example of your search engine on the right-hand side of the dashboard. You can test there if you want, or simply click Save and test back in the simple HTML page. Your changes are shown immediately (figure 7.26).

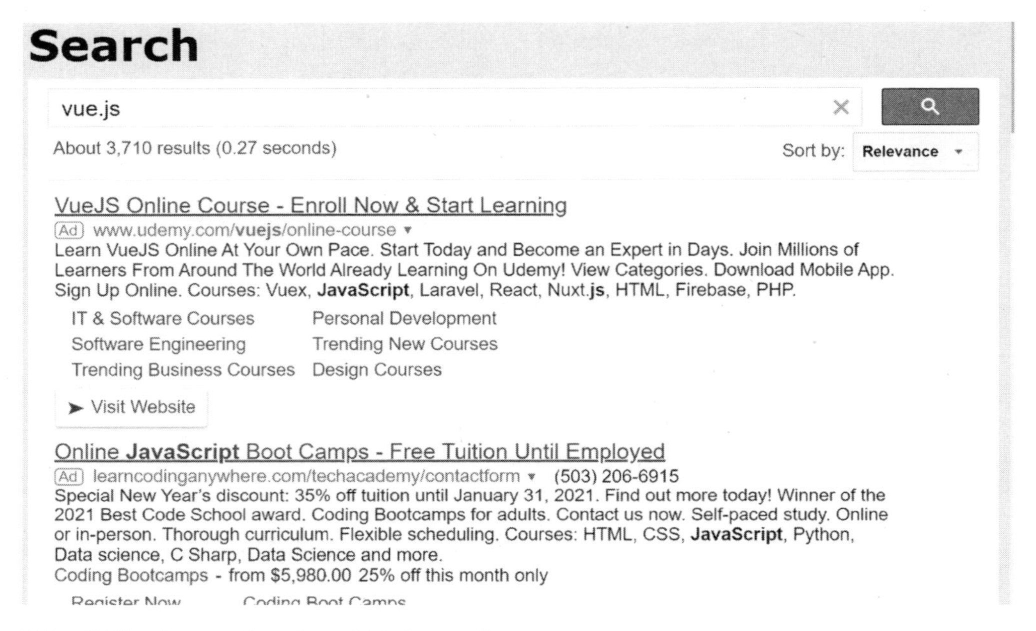

Figure 7.26 The search engine with inline results

Let's make another tweak. If you click on one of the results, you'll see that it opens in a new tab. While that's certainly something you may want, let's change that to load in the current tab instead. In the dashboard, go to "Search features," then "Advanced," then "Websearch settings." In the Link Target field enter "_self" (figure 7.27).

As before, save and reload your test HTML file and click on a result. It should load in the same tab.

Google's Programmable Search Engine is even more customizable if you want to modify the HTML of the snippet. The Developer Guide (https://developers.google.com/custom-search/docs/overview) goes into detail on what can be accomplished. You can also make use of an API that returns results in pure JSON, but this is a commercial feature only. Now that we've considered a plug-and-play solution, let's again look at a more integrated option.

7.2.1 Searching with Lunr

In the previous search example, the search service itself was entirely handled by a third party. You can most likely count on Google being around for a few more years, but what if you want an entirely self-hosted, self-contained solution? One example of this is Lunr (https://lunrjs.com/) (figure 7.28).

Programmable Search

New search engine
Edit search engine
JamstackSearch1
Setup
Look and feel
Search features
Statistics and Logs
Help
Visit Help Forum (Ask a question)
Send Feedback

Promotions | Refinements | Autocomplete | Synonyms | Advanced

Enable or disable advanced search features.

Results sorting — ON

Websearch Settings

Refinement Style — Links / TAB
Results Browsing History — Enable / Disable
Structured Data in Results — Enable / Disable
No Results String
Link Target — _self
Query Addition
Query Parameter Name

Save

Figure 7.27 Modifying search results behavior via the dashboard

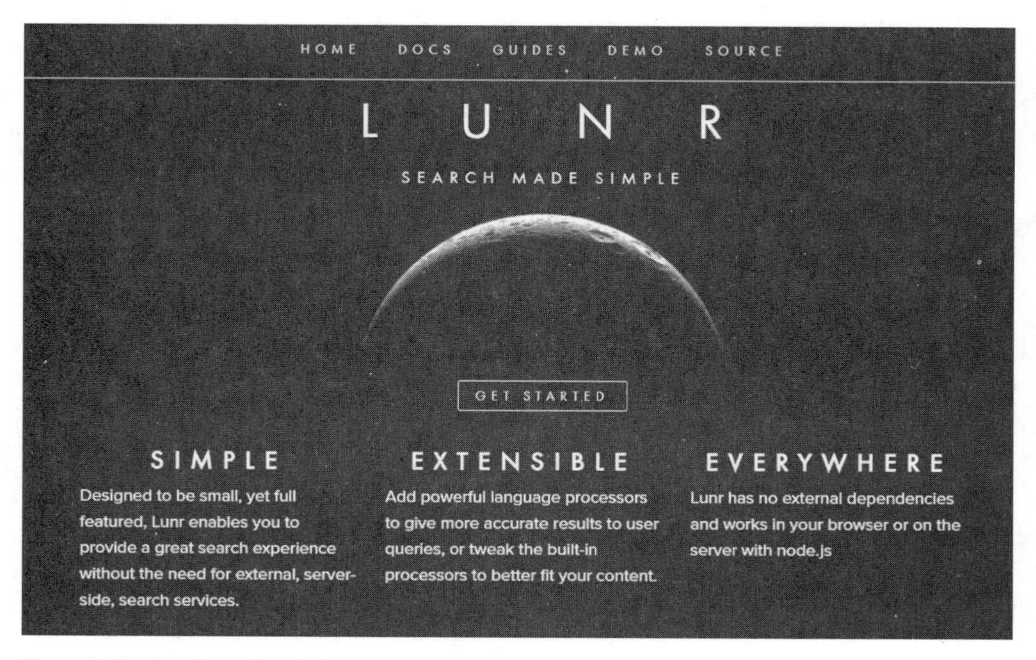

Figure 7.28 Lunr's gloriously simple web page

Lunr is an entirely client-side solution (it can be used server side, but in general this isn't the typical use case) for adding searches to a website. It begins with the creation of an index. This is where you decide what you want to search. For example, a developer service website may contain a few marketing pages and then a deep set of documentation. While you could create an index of everything, you may want to instead focus your index just on the developer documentation.

Even after you've figured out what to search, you still need to consider the size of your index and how you can keep it smaller. Going back to the previous example, if your developer documentation covers hundreds of pages of docs, perhaps you may consider only indexing the first paragraph of each page of docs.

Once you've figured out what content you want to index, you give this information to Lunr. Lunr will parse the text and do magic with it. Okay, not magic, but it will parse the content to make it much more searchable. Once the index is created, you can then search against it using either a simple search term or perhaps a more complex search query ("find all docs that mention cats that are in the API category"). The results can then be filtered by their quality (Lunr will score every result), if you wish. Let's look at how Lunr could be added to a Jamstack site.

ADDING LUNR TO AN ELEVENTY SITE

In chapter 2, you were introduced to the static site generator Eleventy (https://www.11ty.dev/). To test adding search to the Jamstack, we're going to begin with an existing Eleventy site. This site makes use of text from the GI Joe wiki (http://mng.bz/laAB), an online resource for everything related to the GI Joe franchise. I copied part of the description of six different characters to build a very basic site of GI Joe personalities (figure 7.29).

Figure 7.29 The site's home page, which just lists characters

Clicking an individual character link shows some basic text and a picture (figure 7.30).

Cobra Commander

Not much is known of the background of the man many call the Cobra Commander. What we can only tell is how he works and how he thinks. We know that he has deliberately started political and social conflict in a number of areas. He has amassed an army by recruiting displaced people, promising them money, power and a chance to get back at the world that hurt them. In return, he demands that they swear absolute loyalty to his cause. What is his cause? World domination.

Faction: Cobra

Cobra Commander does not delude himself by justifying his actions as matters of principles or glory. He knows he does it for control and power. He is ruthless, hatred-personified and totally obsessed. A meticulous person, he likes to personally oversee vital projects himself, even engaging in military combat himself on occasion. Not much is known about him, he is a master of disguise and he has appeared as a goatee artist looking man with a son in a coma, in the Marvel comics. His appearance in the 12 inch G.I. Joe line shows him as a man with dark slicked back hair, his appearance constantly changing leaves him assumed to wear masks, even the commander can keep his identity from the people around him.

Home -Search

All content sourced from Joepedia.

Figure 7.30 An individual character page

The initial code to create this site consists of a few basic pages. (Remember that you can find this in the book's GitHub repository.) First, the home page loads in all the characters and displays them in a list.

Listing 7.4 Home page for the site (/chapter7/search/lunr/index.liquid)

```
---
layout: main
---

<h1>Characters</h1>

<ul>
{% for character in collections.characters %}
  <li><a href="{{ character.url }}">{{ character.data.title }}</a></li>
{% endfor %}
</ul>
```

One of the ways Eleventy creates collections of related content is by using tags. You can see the source for one of the characters. In the front matter on top, the tag's value is set to characters. This is repeated throughout all the characters. (Note that we've trimmed some of the text to save space.)

Listing 7.5 Example character page (/chapter7/search/lunr/characters/destro.md)

```
---
layout: character
title: Destro
faction: Cobra
tags: characters
image:
➥ https://vignette.wikia.nocookie.net/gijoe/images/c/ca/RAH_Destro02.jpg/
➥ revision/latest?cb=20080730134919
---

Destro is one of the most cunning foes the Joe Team has ever faced. He is
the power behind M.A.R.S. (Military Armament Research Syndicate), one of
the largest manufacturer of state-of-the-art weaponries. His business is
fueled by inciting unstable countries to wage wars against each other and
then getting them to purchase weapons from him. To him, war is simply man's
expression of his most natural state. It is the perfect example of where the
fittest survive and where many technological advances are made. His biggest
client, thus far, is Cobra with whom he maintains an alliance of
convenience. Despite being a manipulative person, Destro maintains a sense
of honor and actually respects the Joe Team for their skills and
expertise, if not their motivation.
```

The templates used to display characters and the home page itself can both be found in the GitHub repository inside chapter7/search/lunr/_includes. These files handle the header and footer and display of the character but don't relate to the search feature we're about to add. We will, however, show the main template shortly when we add support for the client-side aspects of Lunr.

As stated earlier, Lunr requires you to create an index of your data. This is a data-friendly view of the content on your site that you want to search. Our site consists of a home page, multiple character pages, and, eventually, a search page. But we only want to search our character data. In listing 7.4, you saw that you can loop over that information in a collection. We can use the same logic to create a JSON version of the site that Lunr can eventually use to create an index, shown next.

Listing 7.6 Creating a JSON list of characters (/chapter7/search/lunr/lunr.liquid)

```
---
permalink: /index.json    <-- Specifying the destination of the template output
---

[
{% for character in collections.characters %}
{
    "title":"{{character.data.title}}",
    "url":"{{character.url}}",
    "content":"{{ character.templateContent | json }}"    <-- Using a filter to output the content in a JSON-safe way
} {% if forloop.last == false %},{% endif %}    <-- Including a comma for every item except the last
{% endfor %}
]
```

The first thing to note in this template is the use of `permalink` in the front matter. This tells Eleventy to change its normal file-naming behavior for this template and to store the result in a file in the root of the site named index.json. In the template, we use Liquid to dynamically output an array of characters. For each character we output the title, the URL to the character (Eleventy provides this for us in the collection), and then the content. Note that we use a filter, `json`, to manipulate the output. The `templateContent` value provided by Eleventy includes the rendered content of the character. That includes HTML and line breaks, neither of which we want. We'll create that filter shortly. Lastly, notice that we use a Liquid feature of loops that detects if we're on the last iteration. We do this because we need a comma between each item in the array but do not want one after the last item. It may help to see how this JSON looks when done.

Listing 7.7 The generated JSON file

```
[

{
    "title":"Baroness",
    "url":"/characters/baroness/",
    "content":"The Baroness serves as..."
} ,

{
    "title":"Cobra Commander",
    "url":"/characters/cobra_commander/",
    "content":"Not much is known of the ..."
}


]
```

In order to support the JSON safe filter used in listing 7.6, we have to define this in the Eleventy configuration file.

Listing 7.8 The Eleventy configuration (/chapter7/search/lunr/.eleventy.js)

```
module.exports = function(eleventyConfig) {

    eleventyConfig.addPassthroughCopy("css");

    eleventyConfig.addFilter("json", value => {
        //remove html and line breaks
        return value.replace(/<.*?>/g, '').replace(/\n/g,'');
    })
};
```

This line simply tells Eleventy to copy over the CSS used for this simple site.

The filter works by first removing any and all HTML and then line breaks. Note that other methods could be used here as well. For example, what if the character description was incredibly long? Or what if we had way more than six characters? In order to keep the size of the JSON smaller, you could make the call to only return the first thousand letters of the character's description. In the end, Lunr doesn't care, but the more data you send to it the more it will have to work!

Now that we have a data file to use as our source, it's time to build the actual search engine. Figure 7.31 shows how this is going to look. We've got a simple field on top, a button, and that's it, at least initially.

Figure 7.31 The initial search form

After entering some input and clicking Search, the results with links to respective pages will then be displayed (figure 7.32).

Figure 7.32 Displaying search results below the form

The following listing shows how this is built.

Listing 7.9 The search page (/chapter7/search/lunr/search.liquid)

```
---
layout: main
---

<h1>Search</h1>

<input type="search" id="search">
<button id="searchBtn">Search</button>
```

```
<div id="results">
</div>

<script src="https://unpkg.com/lunr/lunr.js"></script>      <-- Load the Lunr library.
<script>
document.addEventListener('DOMContentLoaded', init, false);
let searchField, searchButton, resultsDiv;
let docs, idx;

async function init() {      <-- This code runs when the page loads in the browser.
    searchField = document.querySelector('#search');
    searchButton = document.querySelector('#searchBtn');
    resultsDiv = document.querySelector('#results');

    let result = await fetch('/index.json');
    docs = await result.json();

    // assign an ID so it's easier to look up later,
    // it will be the same as index
    idx = lunr(function () {      <-- This is where the Lunr index is defined.
        this.ref('id');
        this.field('title');
        this.field('content');

        docs.forEach(function (doc, idx) {
            doc.id = idx;
            this.add(doc);
        }, this);
    });

    searchButton.addEventListener('click', search, false);
}

function search() {      <-- This handles user input and performs the search.
    let term = searchField.value;
    if(!term) return;
    console.log('search for '+term);

    let results = idx.search(term);

    console.log(results);

    // we need to add title, url from ref
    results.forEach(r => {
        r.title = docs[r.ref].title;
        r.url = docs[r.ref].url;
    });

    let result = '<h2>Results</h2><ul>';
    results.forEach(r => {
        result += `
<li><a href="${r.url}">${r.title}</a></li>
        `;
    });
```

```
    result += '</ul>';

    resultsDiv.innerHTML = result;
}
</script>
```

The template begins with a very short bit of HTML. We've got the search field, the button, and then an empty div block beneath them. This block will be used to render results. The bulk of the listing is the JavaScript.

Before we get into the JavaScript, we have to add support for Lunr itself. This is done by pointing to the library on a CDN at unpkg.com. We could have also downloaded the library and placed it in our site.

Our code begins by specifying a function to run (`init`) when the page has loaded. Some globally used variables are also defined here to be used later.

The `init` function does a variety of things. First, it creates pointers to the things in the DOM we need to work with: the search field, button, and empty div.

Next, it uses the `fetch` API to load the JSON we defined earlier. It loads this and converts to JavaScript data by parsing the JSON. Once we have that, we can create the Lunr index. We begin by defining the primary key for our data. This is a unique identifier for every item in the index. We've named this "id." Then we then define what's going to be searchable. That will be the `title` and `content` fields from our data.

At this point, the index behavior is defined, but isn't actually filled with anything. To do that, we loop over each item from our JSON array and add it to the index. To create the primary key, we simply use the loop index. This is added to our doc object, which is then added to the index.

The final thing done in the `init` function is adding an event handler for the Search button. When users click that button, we first see if they typed anything, and, if not, simply exit the `search` function.

Searching against Lunr is incredibly easy; it's literally one line:

```
let results = idx.search(term);
```

This provides us an array of results if anything matched. Note the use of console.log immediately after. This lets us use the browser's developer tools to examine the results. This is a handy way to figure out how to work with what Lunr found (figure 7.33).

Now that we have results, we've got to work with them so that we can display what was found. One thing that may be surprising about Lunr is that the search results do not actually contain the document itself. In figure 7.33, you can see that a `ref` value exists. Remember when we defined a primary key for our content and manually added the loop index? This is how we'll be able to display the results. Lunr's use of `ref` basically requires you to associate the result with the original data. This is what the loop is doing after the search. For each result, we add in the title and URL from the original set of documents. With that in place, we can then display the results using some simple HTML.

Search

ninja Search

Home -Search

All content sourced from Joepedia.

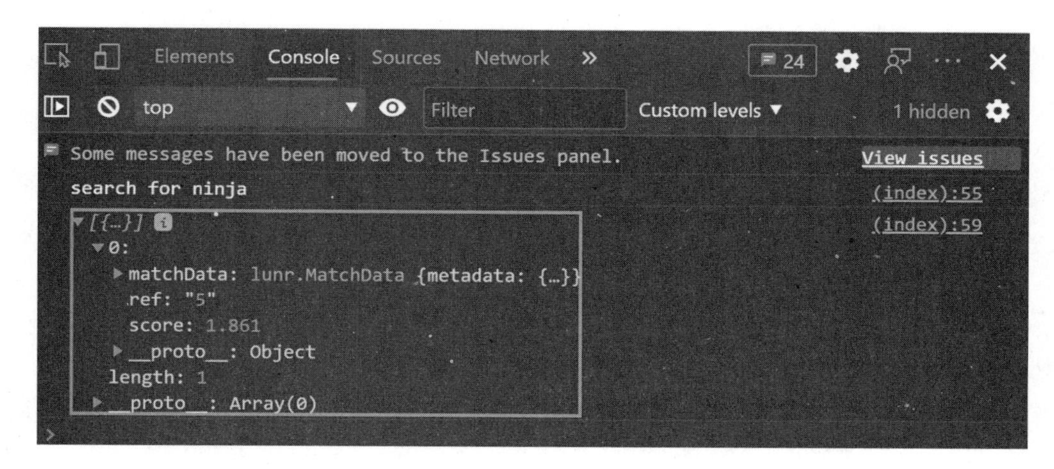

Figure 7.33 Browser dev tools showing the Lunr search results

There is, of course, more that you can do in this template. We could add support for letting the user know nothing was found. We could add support for searching against one particular part of our content (GI Joe characters are divided in teams, and our search interface could let you specify one to search against). Lunr is very flexible in this regard and can support pretty much anything your site needs.

7.2.2 Other options

As with forms, you've got many other options to consider when adding search. One in particular stands out that may be of interest to Jamstack developers: Algolia (https://www.algolia.com/). Algolia is a commercial service (with a generous free tier) that uses an index like Lunr. Unlike Lunr, Algolia hosts your index on its servers and provides API to edit that index and search against it. It also provides powerful analytics about how your visitors search on your site.

If you prefer something similar to Google's service, but would rather not use Google, Microsoft has similar services with Bing Custom Search (https://www.customsearch.ai/). Like Google, it also provides a free version of this service.

7.3 Other dynamic options

In this chapter, we focused on two particular ways of adding dynamic content back into your Jamstack site. Obviously there are a multitude of other ways you can do that. In chapter 5, you saw how to add e-commerce to your website. In the next chapter, you'll see how to do nearly anything by adding serverless features. Some other forms of dynamic content include the following:

- *Calendars*—Google Calendar (http://calendar.google.com/) lets you embed full calendars into an HTML page. Like other Google services, you get some basic styling options. You can also use open source libraries like FullCalendar (https://fullcalendar.io/) to provide completely unique designs to your calendars, which still can be driven by Google Calendar data.
- *Comments*—While typically only used on blogs, you may wish to add comments to your Jamstack site. Probably the most well known is Disqus (https://disqus.com/). This is a pretty standard utility used for commenting across the internet and can be used for free. Other options include Commento (commento.io) and FastComments (fastcomments.com) if you are willing to roll your own solutions, but be prepared to do a bit of work. An article by Matt Mink (https://css-tricks.com/jamstack-comments/) describes just such a system.
- *Chat*—Another common tool on sites (typically commercial sites) is a chat box. This is usually a little talk balloon in the lower right of the site that lets a user click to speak to a person managing a site. A commercial example of such a service is LiveChat (https://www.livechat.com/). Such services can be programmed with canned responses ("When asked about foo, respond with goo"), as well as be connected to living, breathing humans.
- *APIs*—The A in Jamstack stands for API, which means any remote resource (weather, stock data, etc.) that can be called with JavaScript and can be called within your Jamstack site.

Summary

- A site that is simple static HTML, CSS, and JavaScript can still contain dynamic elements (forms, search, calendars, and more).
- Many services exist that make it incredibly easy to add dynamic aspects into your static site. For example, Google has multiple services (forms, search, calendars) where all you need to do is copy some HTML and paste it into your template. Services like WuFoo, FormCake, and FormKeep are just a few options for working with form input, one example of dynamic support.
- Determining which service to use will depend on your needs, your budget, and which best provides the result you need.

8 Working with serverless computing

This chapter covers

- Defining serverless
- Explaining what serverless means for the Jamstack
- Describing serverless options
- Building serverless functions with Netlify
- Building serverless functions with Vercel

In the last chapter, we discussed a few different ways you can add dynamic elements back into your Jamstack sites. These were done via external services set up specifically to serve a need, for example, Google Forms letting you design and accept feedback for a form.

While the availability of these services continues to grow in both number and variety, there are certain things that simply do not make sense as external services, including very particular business logic for your Jamstack site that only you know how to develop. This is where serverless computing, and specifically functions as a service, come into play.

8.1 What is serverless computing?

Let's get this out of the way. There are still servers in serverless computing. Just like how we don't actually build infrastructure in real clouds, serverless technology doesn't magically remove the need for physical hardware. Serverless computing is about removing the need to *worry* about the servers. It's about getting all the benefit of what the server provides without the boring work of actually maintaining a server.

There are a wide variety of services that can fall under the serverless definition. Here are just a few:

- The ability to provision and use a MongoDB database entirely from its website (mongodb.com) without having to install anything themselves.
- The ability to build complex workflows driven by events at Pipedream (pipedream.com). Pipedream lets developers build workflows by piecing together various steps that can include custom code.
- And of course, the ability to build nearly anything with the first real enterprise-level provider of serverless, Amazon Lambda (https://aws.amazon.com/lambda/).

While serverless computing has many interpretations and implementations, this chapter will focus on probably the most widely known and used example: function as a service (FaaS). FaaS lets you focus on the business logic of a particular feature and not worry about the hosting or routing. Let's consider a simple example.

Imagine you want to build a service that returns the current time. First, don't do this. Browsers know what time it is. But this gives us a simple requirement to build. If I were building this with Node.js, the setup would go something like this:

1. Provision a Linux VM.
2. Ensure Node is available.
3. Write a Node script that will listen for requests on a port and particular path.
4. Ensure the VM is set up to respond to that HTTP request.
5. When that happens, perform the business logic (get the current time).

Creating the same result with FaaS would typically be

1. Perform the business logic.

And that's it. Obviously this is a somewhat contrived example, but the idea is that the developer no longer needs to worry about the server, the operating system, the network routing, and such, and only needs to worry about the actual logic of what they're building. It's not always that easy, but you get the idea.

So how does this impact the Jamstack? While there are multiple services out there that provide APIs for various services, sometimes you need very specific logic that only you can create. By using serverless computing and functions as a service, a developer can write precisely what they need, expose it as an API, and then call it with JavaScript from their Jamstack site. Let's look at one form of using this: Netlify functions.

8.2 Building serverless functions with Netlify

We discussed Netlify in an earlier chapter, and one of the features we hinted at was the ability to write serverless functions. This feature, called Netlify functions (https://functions.netlify.com/), lets you include functions written in JavaScript or Go. Netlify handles taking your code and making it available via HTTP, with no additional work required on your part. Netlify supports these functions on their free tier (with caps) and has higher tiers for higher cost (https://www.netlify.com/pricing/#add-ons-functions). Netlify functions are built on Amazon Lambda, but you don't have to know anything about that service in order to use them.

Netlify functions support URL and form parameters, which means your client-side code can pass arguments in multiple ways when executing calls. Your code can also access full details of the request and return data in various forms, although typically you'll return JSON.

For our first test, we're going to make use of the Netlify CLI. While not required, it's going to be useful in this chapter to help us more quickly get started and more easily test our code. To install the Netlify CLI, use npm:

```
npm install -g netlify-cli
```

After installing the utility, you will then be able to run the `netlify` function from your terminal (figure 8.1).

```
ray@Mandalore:~/projects/the-jamstack-book$ netlify

• Netlify CLI
Read the docs: https://www.netlify.com/docs/cli
Support and bugs: https://github.com/netlify/cli/issues

Netlify command line tool

VERSION
  netlify-cli/3.10.6 wsl-x64 node-v14.11.0

USAGE
  $ netlify [COMMAND]

COMMANDS
  addons     (Beta) Manage Netlify Add-ons
  api        Run any Netlify API method
  build      (Beta) Build on your local machine
  deploy     Create a new deploy from the contents of a folder
  dev        Local dev server
  env        (Beta) Control environment variables for the current site
  functions  Manage netlify functions
  help       display help for netlify
  init       Configure continuous deployment for a new or existing site
  link       Link a local repo or project folder to an existing site on Netlify
  lm         Handle Netlify Large Media operations
  login      Login to your Netlify account
  open       Open settings for the site linked to the current folder
  plugins    list installed plugins
  sites      Handle various site operations
  status     Print status information
  switch     Switch your active Netlify account
  unlink     Unlink a local folder from a Netlify site
  watch      Watch for site deploy to finish
```

Figure 8.1 The default output of the Netlify CLI gives basic help for its commands.

After confirming you've installed the CLI correctly, run `netlify login`. This will prompt you for your authentication, so if you didn't create an account with them in the deployment chapter, do so now. Once logged in you won't need to reauthenticate in the future.

As you can see in figure 8.1, the CLI does quite a bit, but we're going to touch on just a few aspects of it in this chapter. In particular, the CLI can scaffold functions for us, including numerous sample apps to help get started.

Your Netlify functions can be stored in any folder you want, but recently Netlify started supporting a default location of netlify/functions. Unfortunately, at the time this book was written, the CLI doesn't yet support knowing about this default, which means we need to specify it. While this can be done via Netlify's web-based dashboard, we can use yet another new feature of Netlify, *file-based configuration.*

Netlify lets you specify a netlify.toml file in the root of your project. Pretty much every setting possible can be set here (see the documentation at http://mng.bz/XWM6 for full details), but in our case we're going to specify only the root folder for our Netlify functions. To get started with our first test, create a new empty directory and add a blank netlify.toml file (or use the chapter8/test1 folder from the GitHub repository). We will assume the folder you created is named `test1`.

Listing 8.1 Configuring netlify.toml (chapter8/test1/netlify.toml)

```
[build]
    functions = "netlify/functions"
```

The configuration file specifies the functions setting in the build group and sets where Netlify should look when preparing to load functions. Again, this value is now the default in Netlify but isn't recognized in the CLI.

At this point, you can create the functions folder if you wish, but the CLI can create it if it needs to. If you decide to create it yourself, you should end up with a structure like the one shown in figure 8.2.

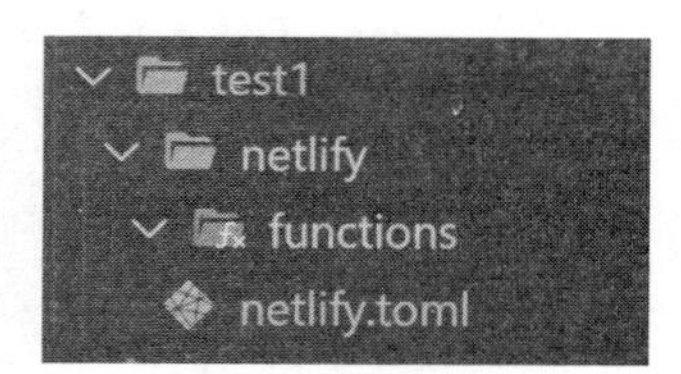

Figure 8.2 Folder structure of our functions test project. Note that netlify.toml is at the root of test1; it's a bit unclear in the screenshot.

Now we can finally test the scaffolding feature of the CLI. To begin, run `netlify functions:create` from the test1 folder. This will open a prompt asking you to select an example (figure 8.3).

```
ray@Mandalore:~/.../the-jamstack-book/chapter8/test1$ netlify functions:create
? Pick a template (Use arrow keys or type to search)
  ----[JS]----
> [hello-world] Basic function that shows async/await usage, and response formatting
  [apollo-graphql] GraphQL function using Apollo-Server-Lambda!
  [apollo-graphql-rest] GraphQL function to wrap REST API using apollo-server-lambda and apollo-datasource-rest!
  [auth-fetch] Use `node-fetch` library and Netlify Identity to access APIs
  [create-user] Programmatically create a Netlify Identity user by invoking a function
  [fauna-crud] CRUD function using Fauna DB
(Move up and down to reveal more choices)
```

Figure 8.3 Browsing the selection of function examples

The first and default option, `hello-world`, is the simplest and the best to start with, so select that first. Next, you'll be prompted to name your function; since we're testing, take the default of `hello-world`. Click Enter, and the CLI will report on the newly scaffolded function (figure 8.4).

```
ray@Mandalore:~/.../the-jamstack-book/chapter8/test1$ netlify functions:create
? Pick a template js-hello-world
? name your function:  hello-world
◇ Creating function hello-world
◇ Created /home/ray/projects/the-jamstack-book/chapter8/test1/netlify/functions/hello-world/hello-world.js
ray@Mandalore:~/.../the-jamstack-book/chapter8/test1$
```

Figure 8.4 The CLI has finished scaffolding the function.

We'll talk more about the general form of Netlify functions later, but for now know that this function looks for a name value in the query string, and if it doesn't exist, uses a default value of `'World'`. It then returns a JSON object with one value message that simply says hello to the name value (again, that is either provided in the query string or defaulted).

Listing 8.2 The `hello-world` function

```
const handler = async (event) => {
  try {
    const subject = event.queryStringParameters.name || 'World'
    return {
      statusCode: 200,
      body: JSON.stringify({ message: `Hello ${subject}` }),
      // // more keys you can return:
      // headers: { "headerName": "headerValue", ... },
      // isBase64Encoded: true,
    }
  } catch (error) {
    return { statusCode: 500, body: error.toString() }
  }
}

module.exports = { handler }
```

So how do we test this? Another useful feature of the CLI is the `dev` command. This starts a local web server and lets you test your Jamstack site locally. In your terminal, ensure you can run `netlify dev`, and you should see output, as seen in figure 8.5. Note that you should still be in the same directory as before.

Along with running a local web server, the CLI may also open a tab in your browser. The site currently doesn't have any HTML files, so you may get a "Not Found" message, but we can ignore that for now. To execute Netlify functions, you address them at /.netlify/functions/nameOfFunction. Given that figure 8.5 says our site is running on localhost at port 8888, the full URL to our test function will be: http://localhost:8888/.netlify/functions/hello-world. Notice there is no ".js" at the

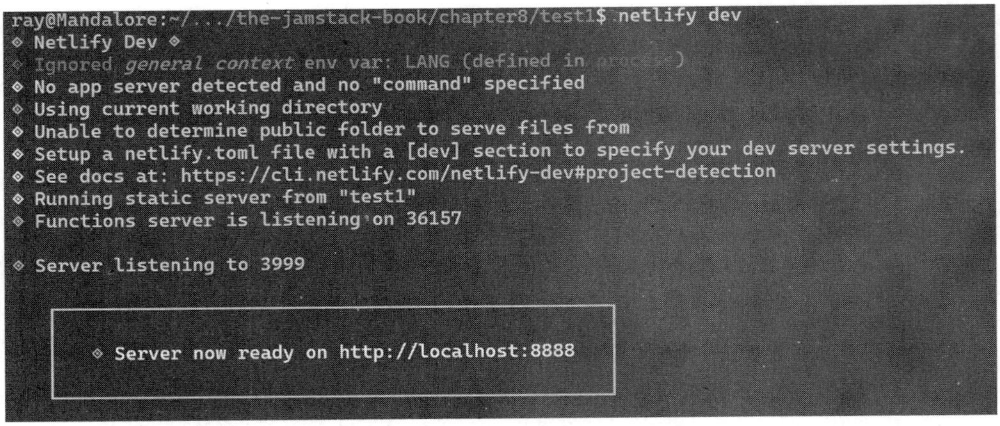

```
ray@Mandalore:~/.../the-jamstack-book/chapter8/test1$ netlify dev
◈ Netlify Dev ◈
◈ Ignored general context env var: LANG (defined in process)
◈ No app server detected and no "command" specified
◈ Using current working directory
◈ Unable to determine public folder to serve files from
◈ Setup a netlify.toml file with a [dev] section to specify your dev server settings.
◈ See docs at: https://cli.netlify.com/netlify-dev#project-detection
◈ Running static server from "test1"
◈ Functions server is listening on 36157

◈ Server listening to 3999

   ◈ Server now ready on http://localhost:8888
```

Figure 8.5 Running the local development server with the Netlify CLI

end. The URL uses the name of the function but not the extension. If you run this, you should get the following output:

```
{"message":"Hello World"}
```

If you remember, we said the function takes a name argument via the URL. You can test this by adding `?name=Ray` (or your own name) to the URL http://localhost:8888/.netlify/functions/hello-world?name=Ray. The output should then update:

```
{"message":"Hello Ray"}
```

Let's build something that makes use of this function. Our site now only contains the function (and configuration file), so let's add an index.html file with a simple JavaScript example.

Best practices dictate that we should (normally) separate our HTML and JavaScript; for this simple demo, one file is sufficient. Our HTML consists of an input field, a button, and an empty div element. The input field will be where users enter their name. The button will fire off a request to our function. And finally, the empty div will display the result.

The JavaScript sets a listener for when the document is loaded, and when it's ready, listens for click events on the button. This runs the function with the somewhat poor name `testApi`, which uses the `fetch` API to hit our Netlify function. If you remember, the result is a JSON object with the key message. We can take the result of that call and write it into our DOM.

Listing 8.3 Using our function from HTML and JavaScript code

```
<!DOCTYPE html>
<html>
<head>
```

```
</head>

<body>

<input type="text" id="name" placeholder="Enter your name...">
<button id="testBtn">Test Function</button>

<div id="result"></div>

<script>
document.addEventListener('DOMContentLoaded', init, false);
let textField, resultDiv;

function init() {
    textField = document.querySelector('#name');
    resultDiv = document.querySelector('#result');
    document.querySelector('#testBtn').addEventListener('click',
        testApi, false);
}

function testApi() {
    let name = textField.value.trim();
    if(!name) return;
    fetch(`/.netlify/functions/hello-world?name=${name}`)
    .then(res => res.json())
    .then(res => {
        resultDiv.innerHTML = `Function returned: ${res.message}`;
    });
}
</script>
</body>
</html>
```

The field where a person can enter their name

Where we listen for clicks on the button

The code executed when the button is clicked. Figure 8.6 demonstrates a simple example of this.

Lindy | Test Function

Function returned: Hello Lindy

Figure 8.6 A test of our Netlify function from simple JavaScript

You now have a simple static site that makes use of a serverless function! Let's look at that serverless function again. At the top, we define the function using arrow-style functions (you can learn more about this style at http://mng.bz/Bx7r). If this style of defining functions is unfamiliar to you, you can rewrite it like so:

```
const handler = async function (event) {
```

Netlify doesn't require arrow functions, but the CLI defaults to using that format. The event object is one of two arguments sent to every Netlify function. The Netlify docs define this as similar to the Amazon AWS API Gateway event, but if you've never used that before, know that it will contain the following values:

- `Path`—The path to the function itself
- `httpMethod`—The HTTP method used to call the function, useful for times when you care if the function was called with a form post

- `headers`—All the request headers for this execution
- `queryStringParameters`—As shown, any values passed along the query string
- `body`—A JSON string of any request payload
- `isBase64Encoded`—A true/false flag specifying if the body is base64 encoded

The second argument passed to Netlify function is a `context` object. This object contains information about the function context itself, for example, things pertaining to the AWS Lambda function behind the scenes (Netlify hides all of this for you!). The context object isn't something you need to use (hence it not even being shown in the function), so you will rarely make actual use of it in your function. The one place it does come into use is with Netlify's user management system called Identity. We aren't covering that here, but you can find out more about that in the docs (https://docs.netlify.com/visitor-access/identity/).

The final part of your function is where you return data. As seen in listing 8.4, an object is returned containing the following parts:

- `statusCode`—This should be a valid HTTP status code representing the result status of the function. 200 is used to mean a good status, whereas 500 (and others) can represent errors and other states. Generally, you will return 200.
- `Body`—A JSON-string representing your data result. The caller will parse this JSON into valid data. You do not need to return a JSON string, but that's fairly typical for serverless functions acting as an API.
- `headers` and `multiValueHeaders`—These let you return headers to be sent along with your data. The `headers` value is used for simple key/value headers (header so-and-so has value so-and-so), whereas `multiValueHeaders` are used in cases where a header has multiple values. Some APIs will use headers to specify the content type or additional data related to things like licensing and other statuses.
- `isBase64Encoded`—Another true/false flag, this time for the result, that specifies if the result is base64 encoded.

Most developers will only need to worry about `statusCode` and `body`, which is what the CLI uses in its default scaffolded function.

Listing 8.4 The `hello-world` function

```
const handler = async (event) => {          // This is the function declaration that defines the event object.
  try {
    const subject = event.queryStringParameters.name || 'World'   // One of the values available in event is queryStringParameters.
    return {                                 // Serverless functions should return a status and body.
      statusCode: 200,
      body: JSON.stringify({ message: `Hello ${subject}` }),
      // // more keys you can return:
      // headers: { "headerName": "headerValue", ... },
      // isBase64Encoded: true,
```

```
    }
  } catch (error) {
    return { statusCode: 500, body: error.toString() }
  }
}

module.exports = { handler }
```

If you want, you can modify your serverless function to log out the entire event object to the console: `console.log(event);`. If you switch to the terminal where you're running the `netlify` command, you can see the output, but note it may be rather large, especially in the request header section, as seen in figure 8.7.

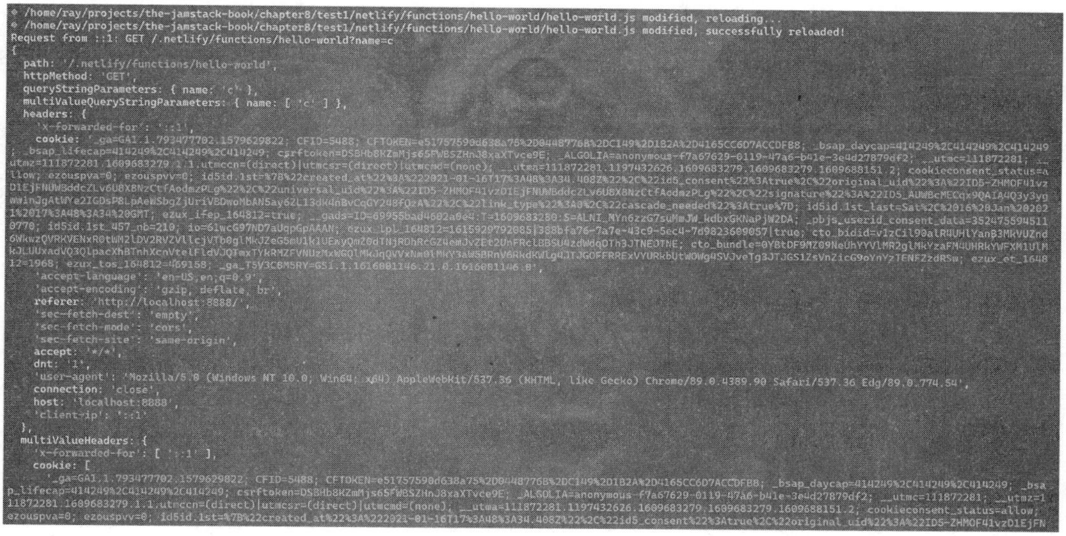

Figure 8.7 Debug output from the serverless function

8.2.1 Adding serverless computing to Camden Grounds

In chapter 2, you learned about Eleventy and built a simple coffee store called Camden Grounds. Let's look at how we can use Netlify functions to enhance our Jamstack site. We're going to modify the site such that each product page will now display a dynamic message if the product is available. We'll use hardcoded logic (everything but tea is available) and client-side JavaScript to display the result. An example is shown in figure 8.8.

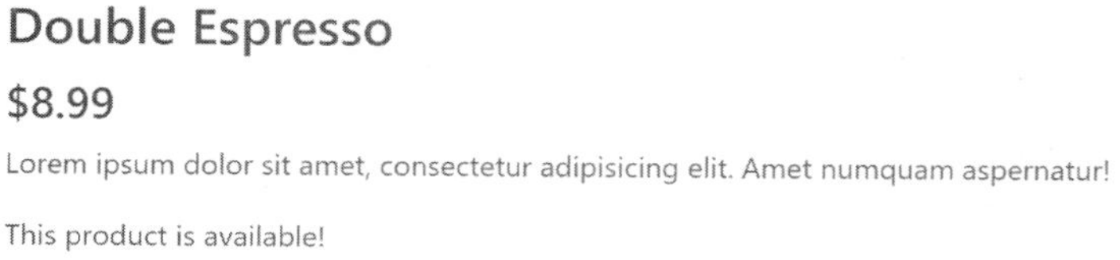

Double Espresso

$8.99

Lorem ipsum dolor sit amet, consectetur adipisicing elit. Amet numquam aspernatur!

This product is available!

Figure 8.8 The double espresso product is available.

Create a new copy of the site into your chapter 8 folder (or ensure you've downloaded the final code from the GitHub repository). Add the netlify.toml file that specifys the function's directory. Since the CLI can create this folder, you do not need to manually create it.

Once again we can use the Netlify CLI to scaffold an application, but this time we'll save a step by specifying the name of function as an argument: `netlify functions:create product-availability`. This should be run in the new folder you created that is the copy of Camden Grounds. You will still be prompted for the type of function to create and accept the default (`hello-world`) again. The result will be a folder named product-availability and a file under it named product-availability.js, as shown in the following listing.

Listing 8.5 The `product-availability` function

```
const handler = async (event) => {
  try {
    const product = event.queryStringParameters.product;    <-- Checking for a product value in the query string
    let available = true;
    if(product.toLowerCase() == 'tea') available = false;   <-- Hardcoded logic that says tea won't be available

    return {
      statusCode: 200,
      body: JSON.stringify({ available }),   <-- Returning the available value
      // // more keys you can return:
      // headers: { "headerName": "headerValue", ... },
      // isBase64Encoded: true,
    }
  } catch (error) {
    return { statusCode: 500, body: error.toString() }
  }
}

module.exports = { handler }
```

As before, we check the query string for a value, this time `product`. If the product being checked is tea, we return `false`. Note the result body uses a shorthand notation that replaces code that would use the same key and value, so, for example, instead of writing `{ name: name }`, we can now write `{ name }` to specify the same thing. Basically, a name key and the value is represented by the name variable. Netlify doesn't require this syntax, but it's there if you want to use it.

To use this, we're going to modify the products.liquid template file, which is used for every product for our site (listing 8.6). Our first change is to add an empty paragraph tag to the product template. This is then edited later by the JavaScript code. This template is executed once for every product, and you can actually see a bit of Liquid in the code itself. While JavaScript code is dynamic in the source, when deployed in product, it's a hardcoded product name.

The JavaScript makes a request to the serverless function, gets the result, and then handles updating the DOM with the availability.

Listing 8.6 The new product template (/chapter8/camdengrounds/products.liquid)

```
---
layout: main
pagination:
    data: products
    size: 1
    alias: product
permalink:"products/{{ product.name | slug }}/index.html"
---

<div class="row my-4">
<div class="col">

     <div class="card mt-4">
          <img class="card-img-top img-fluid"
          src="http://placehold.it/900x400" alt="">
          <div class="card-body">
            <h3 class="card-title">{{ product.name }}</h3>
            <h4>${{product.price}}</h4>
            <p class="card-text">
            {{ product.description }}
            </p>
            <p id="availability"></p>      ◁── A new empty paragraph
          </div>                                that will be updated via
        </div>                                  JavaScript

</div>
</div>

<script>
document.addEventListener('DOMContentLoaded', init, false);
let availabilityP;

async function init() {
  availabilityP = document.querySelector('#availability');
  let api = '/.netlify/functions/product-availability?';
  //add specific product
  api += 'product={{product.name}}';       ◁── JavaScript code that will
  let resp = await fetch(api);                  be static once the site is
  let data = await resp.json();                 deployed
  let result = 'This product is available!';
  if(!data.available)
    result = 'Sorry, this product is <strong>not</strong> available.';
  availabilityP.innerHTML = result;        ◁── Where the actual update happens
}

</script>
```

You've already seen an example of a product being available, but if you go to the Tea page (don't forget to start your site with `netlify dev` if you haven't), you'll now see it isn't available (figure 8.9).

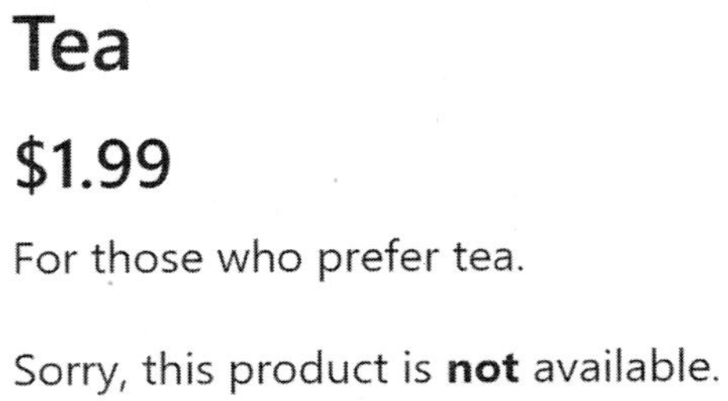

Tea

$1.99

For those who prefer tea.

Sorry, this product is **not** available.

Figure 8.9 No tea for you!

This brings up an interesting situation. Our Camden Grounds site now has both static and dynamic product data. The static portion is in the _data folder and is hardcoded JSON. The dynamic portion is the availability of the product. In production, the list of products will be static, but their availability will be checked on every visit. This makes sense for Camden Grounds, as they rarely change the products they sell but do sometimes run out of a particular product. This would not be appropriate for every Jamstack site out there but shows the flexibility of Jamstack in general.

8.2.2 More on Netlify functions

Netlify functions are both simple to use and powerful in their implementation. You can peruse examples (https://functions.netlify.com/examples/) for more inspiration and play with some live ones at their playground (https://functions.netlify.com/playground/).

Recently, Netlify announced support for "Background Functions" (https://docs.netlify.com/functions/background-functions/). This feature is currently in beta and lets you run functions that need a much longer execution time to process. When called, they return a 202 status immediately and then begin their processing. They would not be used for functions that need to return a result to client-side JavaScript but could be useful for things that need to happen, as you can probably guess, in the background. Again, this feature is currently in beta, so use with caution.

8.3 Building serverless functions with Vercel

Like Netlify, Vercel supports easily deploying and using serverless functions along with your Jamstack site. You can write functions using Node.js, Go, Python, and Ruby. For those who want to use Node.js, you can use either JavaScript or TypeScript, if you prefer. (TypeScript is possible with Netlify functions as well, just not officially supported.) As with Netlify, Vercel has a free tier for their functions; details can be found on their pricing page (https://vercel.com/pricing).

Vercel's serverless support is documented (https://vercel.com/docs/serverless-functions/introduction), but at the time of writing this chapter, there were some parts of that documentation that were a bit misleading. Let's start with a basic introduction to serverless functions with Vercel.

To create a Vercel serverless function, you begin by adding an api folder to your project. (For developers working with Next.js, you would use pages/api instead.) In that folder, you can then create your serverless function. The docs (again, at the time of publication) imply that the name of the function (outside of the extension) does not matter. This is not the case. The name of your file will impact the URL that you use to call the API. Vercel deploys your serverless function at the path /api/filenameMinusExtension. For example, if you create a file named func1.js, and your host is raymondcamden.com, your serverless function would be available at https://raymondcamden.com/api/func1. The documentation demonstrates a path that does *not* include the filename minus the extension, but that will only work if you name your file index.js. In my opinion, it doesn't make sense to use this filename unless you are sure you are going to have only one serverless function on your site. Even if you are only planning one now, in the future you may need more.

This confusion aside, what does a Vercel serverless function look like? Here is the example from its own docs:

```
module.exports = (req, res) => {
  res.json({
    body: req.body,
    query: req.query,
    cookies: req.cookies,
  })
}
```

Once again, an arrow function is being used but is not required. You could rewrite this as such:

```
module.exports = function(req, res) {
  res.json({
    body: req.body,
    query: req.query,
    cookies: req.cookies,
  })
}
```

Use whatever form you are most comfortable with. The function is passed two arguments, `req` and `res`, which map to Node.js `Request` and `Response` objects with additional "helpers" provided by Vercel. For example, req.query maps to the query string of the request. These helpers are documented (http://mng.bz/doOv) and currently exist only for developers using Node.js for their serverless functions. Let's see this in action.

8.3.1 Your first Vercel serverless function

In chapter 6, you tested the Vercel service by installing its CLI tool. If you have not done that yet, please see the instructions in the previous chapter and set that up. You will want to both install the CLI and ensure you run the login command, so you can properly go forward.

Vercel's Zero Config feature means it's rather simple to use most of its features, and that includes its serverless support. In a new folder (or use the /chapter8/test2 folder from the GitHub repository), create an API subdirectory and a new file, func1.js.

Listing 8.7 A Vercel serverless function (/chapter8/test2/api/func1.js)

```
module.exports = (req, res) => {
  res.json({
    body: req.body,
    query: req.query,
    cookies: req.cookies,
  })
}
```

This function is the same code as the Vercel documentation. When requested, it will echo back the request body, query string, and any cookies that were sent. To test this, run `vercel dev` in your terminal, ensuring you are in the folder you created, not the api folder, but rather the directory above it. The `vercel dev` command is much like the Netlify version: it lets you run your site locally and test your site. The first time you run the command, you'll be prompted to answer questions about the site, and you can take all the defaults (figure 8.10).

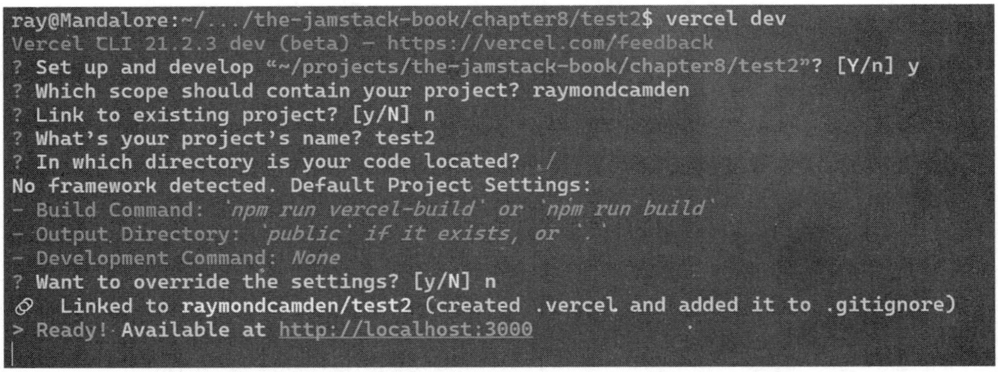

```
ray@Mandalore:~/.../the-jamstack-book/chapter8/test2$ vercel dev
Vercel CLI 21.2.3 dev (beta) — https://vercel.com/feedback
? Set up and develop "~/projects/the-jamstack-book/chapter8/test2"? [Y/n] y
? Which scope should contain your project? raymondcamden
? Link to existing project? [y/N] n
? What's your project's name? test2
? In which directory is your code located? ./
No framework detected. Default Project Settings:
- Build Command: `npm run vercel-build` or `npm run build`
- Output Directory: `public` if it exists, or `.`
- Development Command: None
? Want to override the settings? [y/N] n
🔗  Linked to raymondcamden/test2 (created .vercel and added it to .gitignore)
> Ready! Available at http://localhost:3000
```

Figure 8.10 Running `vercel dev` for the first time

Subsequent runs of `vercel dev` will skip all these questions. To test your function, open your browser to http://localhost:3000/api/func1. The "func1" at the end of the URL is our filename minus the extension. The result will be the contents of the query string (none in our first test), any cookies your browser sent, and nobody, as none was sent in the request. Your cookies may vary, but the output of the query value should look like an empty object:

```
"query":{}
```

If you change the URL in your browser to include a query string, for example, http://localhost:3000/api/func1?msg=hello&name=world, you will see the query value change:

```
"query":{"msg":"hello","name":"world,"}
```

The JSON response you are seeing comes from the serverless function using `res.json`. The argument passed to this is converted to JSON and returned in the result object. You do not need to return JSON, but most serverless functions do.

Let's now test Vercel's serverless functions by replicating what we did in the previous section. Earlier in the chapter, you used the scaffolded Netlify function in a simple frontend application that passed in a name value. Listing 8.4 demonstrated a Netlify function that looks for name in the query string with a default value of `World`. It then returned a JSON object with the key `message` and a value of `"Hello Name"`, where `Name` is either the value in the query string or the default. The new function checks the `req.query` object for a name value and defaults it to `"World"`. Then a string is returned with the `message` key containing the name. As you can see, at least in this example, it's quite a bit slimmer than the Netlify version.

Listing 8.8 `Hello` function with Vercel (/chapter8/test2/api/hello.js)

```
module.exports = (req, res) => {
    let name = req.query.name || "World";

    res.json({
        message:`Hello ${name}`
    })
}
```

Check for a name value in the query string and default it if not present.

If you named this file hello.js, you can then access it at http://localhost:3000/api/hello?name=Ray. If you stopped the Vercel development server, remember to run it again with the `vercel dev` command. After clicking the URL, your output should be:

```
{
  "message": "Hello Ray"
}
```

Now we need to build the simple frontend. This version will be the same as listing 8.3 with one tiny difference: the URL. You can find this new version in the GitHub repo in the file, /chapter8/test2/index.html.

Listing 8.9 Calling our Vercel serverless function (/chapter8/test2/index.html)

```
function testApi() {
    let name = textField.value.trim();
    if(!name) return;
    fetch(`/api/hello?name=${name}`)
```

The changed URL

```
    .then(res => res.json())
    .then(res => {
        resultDiv.innerHTML = `Function returned: ${res.message}`;
    });
}
```

The only thing we changed here is the URL used in the `fetch` command. Both the Netlify and Vercel function returned the same "shape" of their result, which made this update easier. Now let's look at another interesting feature of Vercel's serverless support.

8.3.2 Vercel functions with dynamic path support

In the previous examples, you saw a simple Vercel serverless function using the query string for input. While this works, you can use a different URL syntax that allows for dynamic paths, for example, /api/somefunction/somevalue and /api/somefunction/anothervalue. In these examples, we want to run a function, `somefunction`, and the last values (`somevalue` and `anothervalue`) are inputs given to the function. This makes for slightly simpler URL syntax and could be more desirable than the query string.

To support this, you must do two things. First, create a subdirectory to store your function. This will end up as part of the URL path. Given the example, our directory is /api/somefunction. Next, you create a file for your serverless function. The name of the file must be surrounded in brackets, and the name of the file will determine how you access it in code. For example, if I used [value].js, then my serverless code would access the path value as `req.query.value`. Let's test this by creating a new version of our previous serverless function that lets us use a path and not the query string. Under the api folder, create a new folder, greeter. The name must include the brackets and will then be used by our code to get the value. Create a file named [name].js. The contents of this file are the same as listing 8.8 and can be found in the GitHub repository at /chapter8/test2/api/greeter/[name].js. Once saved, you can access your function at http://localhost:3000/api/greeter/Ray. The output will be the same as the previous example:

```
{
  "message": "Hello Ray"
}
```

If you change Ray to Lindy (http://localhost:3000/api/greeter/Lindy), you can probably guess what you'll get:

```
{
  "message": "Hello Lindy"
}
```

If you want to test this version, the GitHub repository has index2.html, which simply repeats the change to index.html in switching where the `fetch` call makes its request.

8.3.3 Adding serverless functions to Camden Grounds (again)

Let's create another real-world example of serverless functions in the Jamstack by making use of Camden Grounds. This time we're going to add an API to our site. This API will return either all the products available in JSON format or a filtered list based on a query string value.

To begin, copy the Camden Grounds site from your chapter 2 folder, or use the chapter8/camdengrounds2 folder from GitHub. If you don't use the chapter 8 version from GitHub, we have to make a few quick changes to our earlier version. First, remove the existing package.json file. This came from the template source we used and will get in the way with our new version. The easiest way to create a new package .json file from scratch is with the command `npm init -f`. Next, we need to add Eleventy as a dependency. One of the features of Vercel is that it can determine a lot about your project based on your files. The package.json can help tip Vercel off as to how your project works. We've got an Eleventy site, so let's set up Eleventy as a dependency by running `npm install @11ty/eleventy --save`.

At this point, your package.json file should look like so:

```
{
  "name": "camdengrounds2",
  "version": "1.0.0",
  "description": "",
  "main": ".eleventy.js",
  "scripts": {
    "test": "echo \"Error: no test specified\" && exit 1"
  },
  "keywords": [],
  "author": "",
  "license": "ISC",
  "dependencies": {
    "@11ty/eleventy": "^0.12.1"
  }
}
```

Now you can run `vercel dev` to start the local server. As before, you'll be prompted a few times, and you should take the defaults. Once the CLI is done, it will run Eleventy, and this time you'll notice something a bit off in the output (figure 8.11).

What you are seeing is an interesting side effect of using Vercel and Eleventy. When you ran `vercel dev` and it completed its initial setup, it created a new folder called .vercel with some configuration information. You don't have to worry about that. But it also added a .gitignore file that specified .vercel. The .gitignore file is a special file Git repositories use as a way of flagging files and folders that should not be committed to your source repository. This action then triggered something in Eleventy. By default, Eleventy will ignore the node_modules folder used by npm unless you have a .gitignore file. The error message comes from Eleventy *not* ignoring the

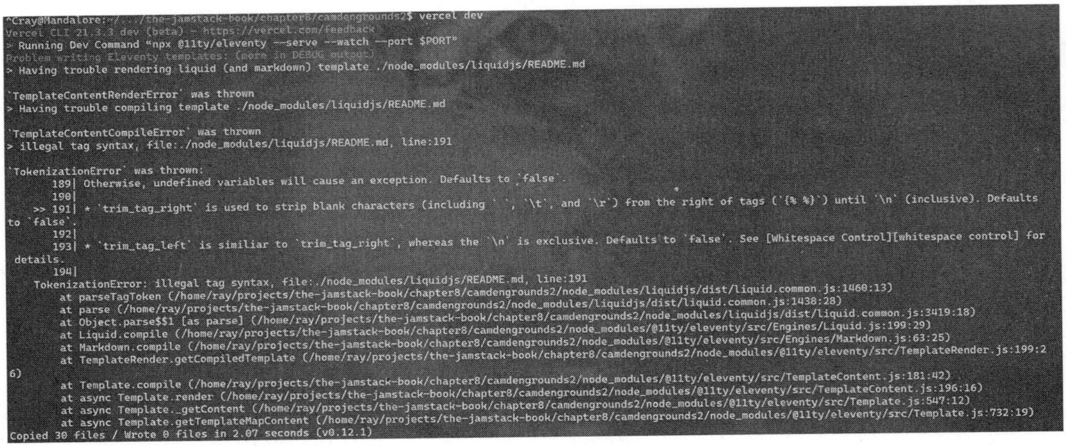

```
^Cray@Mandalore:~/.../the-jamstack-book/chapter8/camdengrounds2$ vercel dev
Vercel CLI 21.3.3 dev (beta) — https://vercel.com/feedback
> Running Dev Command "npx @11ty/eleventy --serve --watch --port $PORT"
Problem writing Eleventy templates: (more in DEBUG output)
> Having trouble rendering liquid (and markdown) template ./node_modules/liquidjs/README.md

`TemplateContentRenderError` was thrown
> Having trouble compiling template ./node_modules/liquidjs/README.md

`TemplateContentCompileError` was thrown
> illegal tag syntax, file:./node_modules/liquidjs/README.md, line:191

`TokenizationError` was thrown:
      189| Otherwise, undefined variables will cause an exception. Defaults to `false`.
      190|
   >> 191| * `trim_tag_right` is used to strip blank characters (including ` `, `\t`, and `\r`) from the right of tags (`{% %}`) until `\n` (inclusive). Defaults
to `false`.
      192|
      193| * `trim_tag_left` is similiar to `trim_tag_right`, whereas the `\n` is exclusive. Defaults to `false`. See [Whitespace Control][whitespace control] for
 details.
      194|
    TokenizationError: illegal tag syntax, file:./node_modules/liquidjs/README.md, line:191
        at parseTagToken (/home/ray/projects/the-jamstack-book/chapter8/camdengrounds2/node_modules/liquidjs/dist/liquid.common.js:1460:13)
        at parse (/home/ray/projects/the-jamstack-book/chapter8/camdengrounds2/node_modules/liquidjs/dist/liquid.common.js:1438:28)
        at Object.parse$$1 [as parse] (/home/ray/projects/the-jamstack-book/chapter8/camdengrounds2/node_modules/liquidjs/dist/liquid.common.js:3419:18)
        at Liquid.compile (/home/ray/projects/the-jamstack-book/chapter8/camdengrounds2/node_modules/@11ty/eleventy/src/Engines/Liquid.js:199:29)
        at Markdown.compile (/home/ray/projects/the-jamstack-book/chapter8/camdengrounds2/node_modules/@11ty/eleventy/src/Engines/Markdown.js:63:25)
        at TemplateRender.getCompiledTemplate (/home/ray/projects/the-jamstack-book/chapter8/camdengrounds2/node_modules/@11ty/eleventy/src/TemplateRender.js:199:2
6)
        at Template.compile (/home/ray/projects/the-jamstack-book/chapter8/camdengrounds2/node_modules/@11ty/eleventy/src/TemplateContent.js:181:42)
        at async Template.render (/home/ray/projects/the-jamstack-book/chapter8/camdengrounds2/node_modules/@11ty/eleventy/src/TemplateContent.js:196:16)
        at async Template._getContent (/home/ray/projects/the-jamstack-book/chapter8/camdengrounds2/node_modules/@11ty/eleventy/src/Template.js:547:12)
        at async Template.getTemplateMapContent (/home/ray/projects/the-jamstack-book/chapter8/camdengrounds2/node_modules/@11ty/eleventy/src/Template.js:732:19)
Copied 30 files / Wrote 0 files in 2.07 seconds (v0.12.1)
```

Figure 8.11 Error output from the Eleventy CLI

node_modules folder and getting tripped up by a random file inside. Luckily the fix is simple: add the node_modules file to your .gitignore file.

Listing 8.10 The fixed .gitignore file (/chapter8/camdengrounds2/.gitignore)

```
.vercel
node_modules
```

At this point you can press Ctrl or CMD-C in your terminal to stop `vercel dev` and simply run it again; the error should go away.

Now Camden Grounds should be up and running. Let's add our serverless function. Create an api folder and then a file named products.js. The file begins with a copy of our product data from the Eleventy _data folder (with some of the text trimmed in the code listing to make it a bit shorter). After the list of products, the meat of the function begins. First, it looks for a query string value named `filter`. If it exists, this is used to filter the array. Finally, the products are returned as JSON.

Listing 8.11 The products API (/chapter8/camdengrounds2/api/products.js)

```
let products = [
    {
        "name" : "Coffee",
        "price" : 2.99,
        "description" : "Lorem ipsum dolor sit amet, consectetur adipisicing
        ➥ elit. Amet numquam aspernatur!",
        "thumbnail" : "http://placehold.it/700x400",
        "image" : "http://placehold.it/900x350"
    },
    {
        "name" : "Espresso",
        "price" : 3.99,
```

```
        "description" : "Lorem ipsum dolor sit amet, consectetur adipisicing
        ➥ elit. Amet numquam aspernatur!",
        "thumbnail" : "http://placehold.it/700x400",
        "image" : "http://placehold.it/900x350"
    },
    {
        "name" : "Americano",
        "price" : 5.99,
        "description" : "Lorem ipsum dolor sit amet, consectetur adipisicing
        ➥ elit. Amet numquam aspernatur!",
        "thumbnail" : "http://placehold.it/700x400",
        "image" : "http://placehold.it/900x350"
    },
    {
        "name" : "Double Espresso",
        "price" : 8.99,
        "description" : "Lorem ipsum dolor sit amet, consectetur adipisicing
        ➥ elit. Amet numquam aspernatur!",
        "thumbnail" : "http://placehold.it/700x400",
        "image" : "http://placehold.it/900x350"
    },
    {
        "name" : "Tea",
        "price" : 1.99,
        "description" : "For those who prefer tea.",
        "thumbnail" : "http://placehold.it/700x400",
        "image" : "http://placehold.it/900x350"
    }
]

module.exports = (req, res) => {
    let filter = req.query.filter;
    if(filter) {
        products = products.filter(
            p => p.name.toLowerCase().indexOf(filter.toLowerCase()) >= 0
        );
    }

    res.json({
        products
    })
}
```

Note that while testing, I noticed that the first time I added my serverless function, the CLI did not recognize that it had been added. I stopped the Vercel CLI and reran, and it picked it up fine. If you encounter this as well, try the same.

With the function saved, access it in your browser at http://localhost:3000/api/products. You'll see all the products. Next, try filtering http://localhost:3000/api/products?filter=espresso. This returns a shorter list:

```
{
  "products": [
    {
      "name": "Espresso",
```

```
      "price": 3.99,
      "description": "Lorem ipsum dolor sit amet, consectetur
     adipisicing...",
      "thumbnail": "http://placehold.it/700x400",
      "image": "http://placehold.it/900x350"
    },
    {
      "name": "Double Espresso",
      "price": 8.99,
      "description": "Lorem ipsum dolor sit amet, consectetur
     adipisicing...",
      "thumbnail": "http://placehold.it/700x400",
      "image": "http://placehold.it/900x350"
    }
  ]
}
```

Now your "static" Jamstack site has its own API that can be used by other sites!

In the next chapter, you'll get an introduction to powerful, user-friendly CMS (content management system) offerings that make Jamstack even easier to use, especially for people without a technical background.

Summary

- Serverless computing gives you all the power of a server without the boring chore of having to set one up. It also relieves you of the worry of maintaining that server.
- FaaS is a generic term that describes using serverless computing as a way of writing functions that perform some action and then exposing that function via a URL. These functions could be site-specific utilities that don't exist as external services. For example, a store that has unusual hours (let's say weather-dependent) could use a serverless function to handle that logic. It could use another serverless function to handle checking product availability.
- Netlify provides support for serverless functions and has a CLI that can be used to scaffold and test locally. Netlify lets you build multiple functions in your Jamstack site and provides a standard URL for accessing them.
- Vercel has support for serverless functions and can be tested locally via the CLI. Like Netlify, it follows a standard of where to place your code and how to access it.

9 Adding a content management system

This chapter covers

- The pros and cons of the two types of headless CMSs
- Choosing between the Contentful and Sanity API-based headless CMSs
- Using WordPress as a headless CMS
- Building a site using Gatsby's WordPress integration
- Using site builders to create Jamstack sites connected to a CMS

Jamstack is often associated with a file-based content management system that uses a combination of manually edited Markdown and YAML files. That's because many static site generators default to this type of manual content management, even today. But the reality is that some form of headless content management system has become a de facto integration in most Jamstack sites.

Jamstack.org maintains a list of headless CMSs (https://jamstack.org/headless-cms) that you can search by type and license (i.e., whether they are open or closed source). Currently, it lists 87 options. Needless to say, that's a lot of choices and more than we can cover in a chapter. Instead, we'll look at the different types of headless CMS available and then explore some options to give you a sense of how their approaches differ and to help you make informed choices about what you are looking for.

9.1 Types of headless CMSs

As we discussed in chapter 4, there are two broad types of headless CMS:

- *Git-based headless CMSs*—These CMSs store everything as files in a Git repository. The CMS is an editing layer that allows the user to edit the Markdown, JSON, and YAML files containing the site's content via a browser-based what-you-see-is-what-you-get (WYSIWYG)-style interface. In many cases, these editing layers are deployed with the site's files or exist locally on the user's machine. The user's changes are committed back to the Git repository by the CMS. Git handles versioning the content. Since changes are committed to Git, deploying content typically happens as part of whatever continuous deployment process was set up for the site. For example, Netlify will detect changes to the repository and automatically build and deploy the updated content.
- *API-based headless CMSs*—These CMSs store content within their systems. This is similar to a traditional CMS, but the system does not have a frontend; instead, content is only accessible via an API. Content editing is typically performed via an admin dashboard on the CMS provider's site. For example, if a site used Contentful, a popular API-based headless CMS, the site's content editors would log in at contentful.com to make changes to the content. When a user makes a change to a Jamstack site's content, Jamstack usually needs to trigger a rebuild via a web hook connected to the site's continuous deployment process. The user does not need to wait for the rebuild to complete in order to view and test their changes. Most API-based headless CMSs have a live preview feature that can display unpublished content changes on a Jamstack site built with popular frontend frameworks like React or Vue.

What is a web hook? Build hook? Deploy hook?

A *web hook* is defined as a user-defined callback that is sent via HTTP. Essentially, an application will call an endpoint, frequently passing some data, in response to an event. For example, a headless CMS will often allow you to configure a web hook to be called when a change is made to the published content.

A *build hook* (as Netlify calls it) or *deploy hook* (as Vercel calls it) is a specialized web hook endpoint that can be called via an HTTP POST request, which triggers a rebuild of the specified site on the deployment provider. You can use a combination of a web hook, at your headless CMS provider, and a build hook, at your deployment provider, to trigger a site rebuild when a change is made in the CMS to the published content of the site, thus incorporating the new content into the published site.

Which type should you choose for your project? That depends on your specific needs, but here are some guidelines.

9.1.1 Pros and cons of Git-based headless CMSs

Pros:

- *They're inexpensive.* Since they rely on a Git repository for all the storage and versioning of content, Git-based headless CMSs are usually inexpensive or, in some cases, completely free and/or open source.
- *They're very easy to set up.* Integration frequently consists of ensuring that the editing layer understands the form (i.e., content model) of the content in your files. For example, you may need to tell it what types of data are in the front matter for Markdown files or the type of data in your JSON and YAML data files, although some systems will even attempt to auto-detect these.
- *They have a short learning curve.* Because they are file-based, they are generally easier for content editors to learn, as data models are typically not as complex as those of their API-based alternatives.
- *They offer easy paths for external contributions.* Third-party contributors can get access to editing content directly through the CMS via features such as open authoring in Netlify CMS, for example, or can make direct file edits via the Git repository on a site like GitHub or GitLab. This can be an especially important consideration for things like documentation sites, as we explored in chapter 4.
- *They're not proprietary.* All of your content is stored in files in your own Git repository, meaning that you can easily swap to another Git-based CMS solution without any direct impact on your site's code or content.

Cons:

- *Content reuse is limited.* Content in a Git-based headless CMS is generally specific to the web, meaning content can be difficult to reuse across a website and mobile app, for instance. In addition, content reuse across the site itself is limited. For example, I may have a marketing headline and tagline that gets used in multiple locations on a site, but this kind of reuse is usually not possible with a Git-based CMS.
- *Data models are difficult to enforce.* Since the content is file-based, direct changes to the files in the Git repository can break the data model's integrity, potentially causing the build to fail. For instance, the CMS may require that every blog post have an author defined in the front matter, but this cannot be enforced on direct file changes in the repository. If someone were to manually change the file and remove the author field, the site build would likely break.
- *Content relationships are difficult to maintain.* Most Git-based CMSs have a way of defining the relationship between two pieces of content. For example, a blog post could be related to an author page with an author name and bio. However, there is generally no way to enforce these relationships, meaning that someone could remove the author page even though the posts assigned to it still exist.

9.1.2 Pros and cons of API-based headless CMSs

Pros:

- *They're designed for content reuse.* This means both reusing content across multiple properties, including multiple sites or a site and a mobile app, and across the same site. To enable content reuse across different types of applications (e.g., mobile app and web app), API-based headless CMSs generally separate the content from the presentation (i.e., the content is not stored in HTML or formats like Markdown that only generate HTML). This allows the content to be used across different types of applications or different parts of the same application without worrying about how the content is displayed.
- *They enforce data integrity of the content.* The content model can be designed to ensure not just that a content type always has certain fields, but that those fields contain the proper values. For example, I may want to ensure that a property of a content item is a number but also within a specific number range. The CMS provides tools for defining field validation that ensure the values are correct before a user is able to publish an item.
- *They enforce content relationships.* Enforcing content relationships makes them easy to maintain. For example, it can ensure that a post always has an author and disallow removing an author that is assigned to published posts. The CMS is designed to prevent the removal of content that would damage the integrity of any references. The references would have to be removed or modified before the content could be deleted. They can also allow relationships within rich text or structured content, which are types of content that allow both rich formatting that isn't tied to HTML and embedding of reusable content items defined within the CMS.
- *They can handle a complex content workflow.* Git-based CMS systems usually only support a very basic workflow, but API-based systems can typically handle even complex and custom approval workflows for content.

Cons:

- *They can be expensive.* Many of these solutions are priced for companies, not individual developers. While they may have generous free tiers, the pricing can often escalate quickly.
- *They often have steep learning curves.* Since these are designed to meet the needs of even large enterprises, they are often full of features that can make them a bit complex for the average content editor.
- *They are proprietary.* There is a built-in vendor lock-in, largely because moving from one system to the other can be a complex and expensive project to undertake. This involves not only moving the content from one proprietary system to another, including data models, users, validation rules, and workflow. It also will involve major modifications to the site's code, pulling out the SDK and/or completely overhauling the API calls that get content.

9.2 Exploring popular headless CMSs

Since we discussed the popular Git-based headless CMS in chapter 4, let's explore some of the widely used API-based solutions. We'll also look at some options to get started implementing them in some of the SSGs discussed throughout this book, although a full discussion of how to implement each headless CMS in each specific SSG is outside the scope of this chapter.

> **Comparing headless CMS options**
>
> The list of headless CMSs on Jamstack's website (https://jamstack.org/headless-cms) is a very comprehensive list of headless CMS options, but it may not offer enough detail to help you make a choice. Thankfully, the folks at Bejamas, a Jamstack consultancy, launched a project they call the Discover Hub (https://bejamas.io/discovery). It offers extremely detailed reviews of 24 different headless options (as well as other categories of tools and services, including SSGs and hosting). Definitely check it out when making a choice.

The first step in using any headless CMS is building a content model. Let's look at how you'd build a content model in two different headless CMS options that have two distinctly different approaches in order to give you a sense of the different options available to you.

9.2.1 Contentful

In my opinion, one of the strongest selling points of Contentful (https://www.contentful.com/) is the ease with which you can define a content model. Its web-based visual interface is easy enough to use that the content model could be built by a nontechnical user.

Once you've logged into Contentful, you'll need to create your first "space." This is how Contentful groups projects. In the case of a Jamstack project, one space might be the backend for one site. However, a space can technically serve as the backend to any logical grouping of sites and apps. Keep in mind that Contentful currently provides only one limited space under its free plan.

Once you've created a space, you'll need to add content types. Under the Content Model heading, click the blue Add content type button. This will open a window to give your type a name and an API identifier and description. The name can be anything you like. The API identifier is essentially a slug for your content type and will generate automatically, though you can customize it.

Once you've created a type, you'll need to add fields to it. Every field must be assigned one of nine different types. Within Contentful's system, each type has a limited set of validations that can be applied to them (figure 9.1).

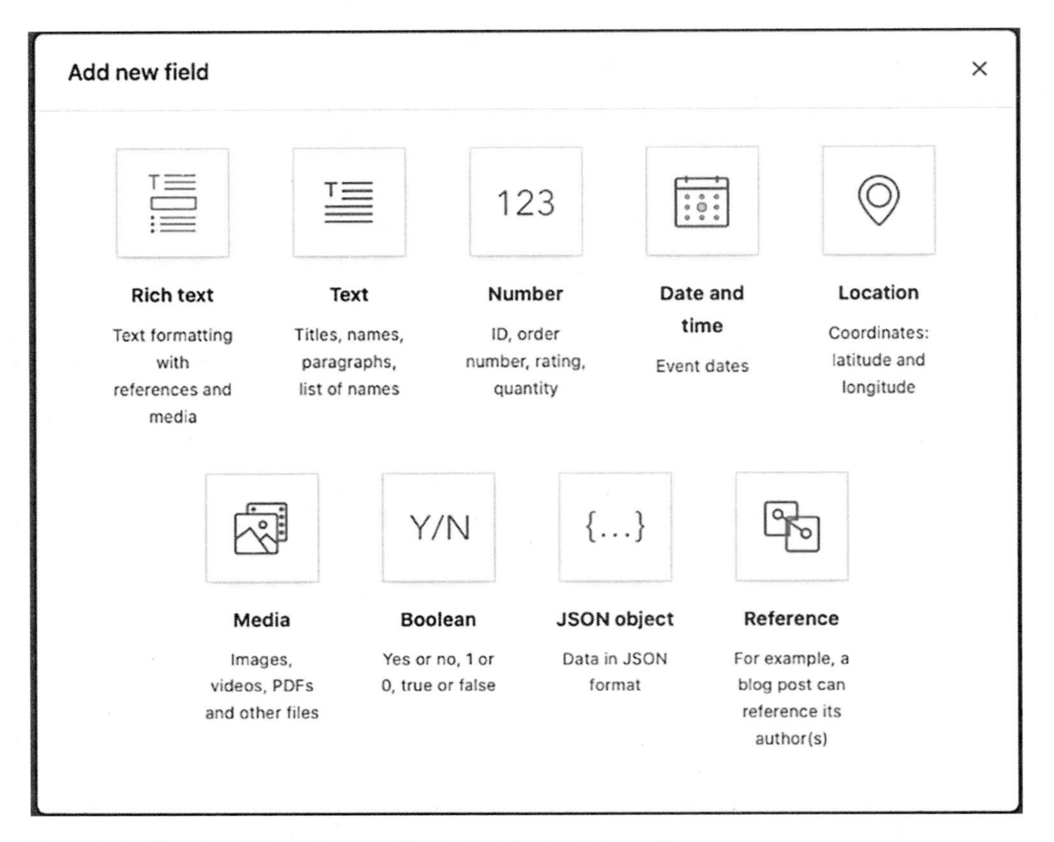

Figure 9.1 The nine different types of fields that Contentful provides

- *Rich text*—It is important to note that this is not HTML, but rather the rich content stored in a JSON format that allows it to be converted to whatever format needed. This allows rich text to include embedded assets and entries within the rich text format. For the content creator, this will look and feel much like a standard WYSIWYG editing experience. There are quite a few validation options beyond simply making it required or limited to a specific character count, all of which relate to embedded links, entries, or assets.
- *Text*—There are two types of standard text fields in Contentful: short text that is limited to 255 characters and long text of up to 50,000 characters. You can sort short-text fields and search for exact matches; while you cannot sort on a long-text field you can do a full-text search. Long text is primarily for the long-form content of an item, such as the blog post body or the author bio (either can be rich text as well). If you are storing Markdown content, you will use a long-text rather than rich-text field. The validation options include things like ensuring the text field matches a specific pattern (e.g., an email or a URL) or prohibiting specific patterns (e.g., preventing the use of swear words).

- *Number*—This can be either an integer or a decimal. Validation can require that the value be unique (no two entries can have the same number), must be within a specific range, or can accept only specific values.
- *Date*—Date does not have many options or validations, although you can limit it to dates within a specific range.
- *Location*—This will be stored as latitude and longitude within Contentful, although content editors will be able to choose a location by address or location on a map. The only validation offered is to make this required.
- *Media*—This is any type of media you'd upload, such as images or PDFs. You can choose to allow a single file or multiple files to be uploaded. Validation allows you to specify file size limits, accept only specific types of files, or limit image uploads to specific dimensions.
- *Boolean*—This is a basic yes/no response. The only validation that can be set is to make it required.
- *JSON object*—This allows the content editor to directly write a JSON object via a JSON editor. JSON validation is built into the editor. You can limit the number of properties the JSON object has via the validation.
- *Reference*—This is a relationship to another content item within Contentful. For instance, a blog post may reference its author or authors. You can have a reference to a single item (a one-to-one relationship) or multiple items (a one-to-many relationship). Validation allows you to limit the option to a specific type (e.g., the author relationship would limit you to choosing content items of the author type).

Once you've built a content model, you and/or your content editors can begin adding content items to Contentful via the content editor, as shown in figure 9.2.

Contentful offers seven different APIs (http://mng.bz/raAe) for retrieving and managing content and assets. For the purposes of pulling data for the frontend of a Jamstack site, you'll be primarily interested in three:

- The Content Delivery API is a REST API that only allows you to retrieve published content. This is the primary API that you'll use to pull content from when generating static pages for the published site. (See http://mng.bz/VlmW for more details.)
- The Content Preview API is a REST API that allows you to pull unpublished changes to content. This is important to allow content editors to view what their changes to content will look like on the live site via a preview version of the site available at a different URL or available to logged-in users on the main site, for example. (See http://mng.bz/xvAg for more details.)
- The GraphQL Content API is a GraphQL API that allows you to retrieve both published and unpublished content. This can take the place of the REST APIs, provided you are comfortable working with GraphQL. (See http://mng.bz/Ax7g for more details.)

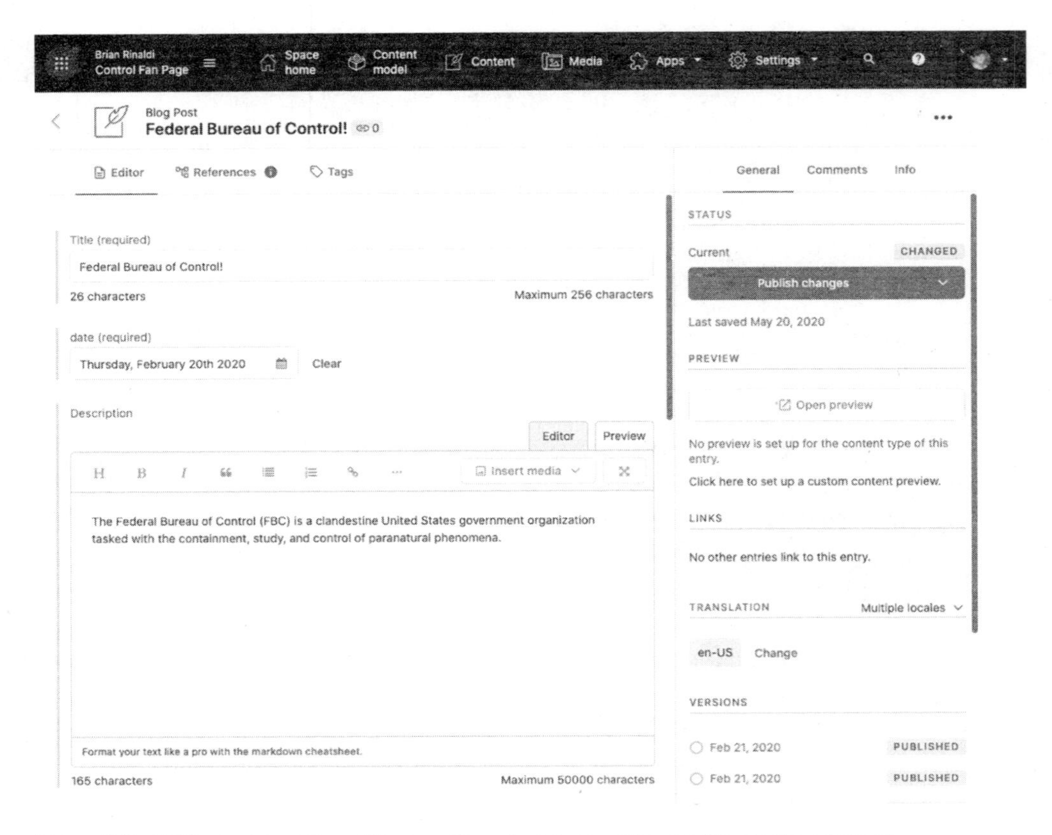

Figure 9.2 Editing content for a "blog post" content type within the Contentful admin UI

In most cases you probably won't be directly interacting with the API and will be using one of Contentful's provided SDKs (http://mng.bz/Zzv5) instead. For instance, you can integrate the JavaScript SDK (http://mng.bz/mxAM) with SSGs like Eleventy or Next.js. Gatsby already has a gatsby-source-contentful (http://mng.bz/5KRD) plug-in that is already designed to work with the GraphQL content API to enable both published and preview content. Next.js has an example that you can use as a starter (http://mng.bz/6ZR6).

Integrating with other traditional SSGs, like Hugo or Jekyll, requires a different approach since neither has a method for directly fetching content from an API nor for generating pages. However, there are tools that can provide ways to import your Contentful content into these tools. Contentful has a jekyll-contentful-data-import (http://mng.bz/oaAv) gem that can be used to import content into Jekyll. For Hugo, there is a contentful-hugo npm package (http://mng.bz/nYA4) that will pull all your Contentful content and convert it to Markdown.

9.2.2 Sanity

The experience of building a content model in Sanity is completely different from that of Contentful. Whereas Contentful primarily relies on its web-based GUI for building content models, Sanity's content models are defined entirely in JavaScript code and managed via the Sanity CLI. (See table 9.1 for a comparison of the content types of Contentful and Sanity.)

Table 9.1 Comparing the different content types available within Contentful and Sanity

Content Type Description	Contentful Type	Sanity Type(s)
Rich text, including HTML formatting, plus embedded content	Rich text	Block
A Boolean or true/false	Boolean	Boolean
Raw text, Markdown, URLs, slugs, JSON	Text JSON	String Text Slug URL
Numbers, including integers and decimals	Number	Number
Dates or dates with time	Date	Date Datetime
A location determined by latitude and longitude	Location	Geopoint
Media, including images, videos, and files	Media	File Image
References to other content objects in the CMS	Reference	Reference Array

Sanity believes that defining content models in code can give you better control over the model while also enabling you to add your own custom validations. Nonetheless, it also means that building and maintaining the content model requires development resources, where alternatives that use a visual editor (like Contentful, for example) might not. In addition, Sanity deploys a new instance of Sanity Studio, its open source, web-based content management interface, with every project. Since you have access to the admin site code for the project, you are free to modify the interface however you choose, but the initial interface has fewer frills than the Contentful interface.

Sanity has 17 types compared to Contentful's 9, but they are largely similar to Contentful's types:

- Array
- Block
- Boolean
- Date
- Datetime
- Document
- File
- Geopoint
- Image
- Number
- Object
- Reference
- Slug
- String
- Span
- Text
- URL

In many cases, where Contentful has subtypes, Sanity has multiple types. For example, Contentful has a single Reference type that refers to one-to-one and one-to-many relationships between content items. In Sanity, there is a Reference type as well, but it only refers to the one-to-one relationship. The one-to-many relationship refers to the Array type. Where Contentful has a single Media content type that refers to all types of media uploads, Sanity has a File and Image type for different types of media uploads.

Sanity also provides a structured text format similar to rich text in Contentful called Block. Sanity's Block type follows the Portable Text specification (https://github.com/portabletext/portabletext) for rich text, meaning it is made up of an array of subtypes that represent the types of content within the block. These can be standard types, like headings, lists, or links, but they can also be custom types. For instance, your content model may have an Employee type, and the block element for an Our Team page would allow you to add items of the Employee type.

Every content type in Sanity has properties such as whether it is `readonly` or `hidden` from the admin UI. All types share the same core properties, although some, like Block and Image, have additional properties that can be set. Each content type also has a set of predefined validations, such as whether the field is required or, in the case of Text, has a minimum, maximum, or specific length requirement. These validations vary per type. Sanity also allows you to specify a custom validation (http://mng.bz/voDr) on any type that is written as a JavaScript function. This means you can create complex validations that are not covered by the built-in validation types.

9.2.3 Defining a content model using Sanity

Since it is entirely code based and the editor UI can be run locally, the experience of building a content model using Sanity can be quite unique. Let's walk through how to create a content model based on a Sanity-provided sample project and explore the sample schema code to give you a better sense of how Sanity works.

To create a Sanity project, you'll need a Sanity account, and then you'll use the Sanity CLI, which can be installed via npm. Then, you'll initialize a new project:

```
npm install -g @sanity/cli
sanity init
```

If you haven't logged in via the CLI yet, the CLI will open a browser window to authenticate you. After that you'll be asked to either choose an existing project to launch, if you have one already, or create a new project. You'll give your project a name and then choose if you want to use the default data set configuration (by default, the production data set can be queried without authentication). Next, you'll choose an output path, which will default to the current project folder. This is where all the local files necessary to configure your Sanity schema and Sanity Studio will be placed. Finally, you can choose either a blank schema or start with a preexisting example:

```
> Movie project (schema + sample data)
  E-commerce (schema + sample data)
  Blog (schema)
  Clean project with no predefined schemas
```

The Movie project schema is a good beginner's example for trying out Sanity, as it has both a simple schema and data to populate it. This offers you the opportunity to try out the studio without needing to manually populate it with data. You'll want to answer yes when asked whether to import the sample data. Once the import is complete, run `sanity start` from the command line to start Sanity Studio. By default, it will run locally on http://localhost:3333.

All the files that make up the content model are in the schemas folder within the project. Let's quickly explore the Person type defined in the person.js file in listing 9.1, and then we'll dig into how it works.

Listing 9.1 The schema file defining a person content type in Sanity

```
import UserIcon from 'part:@sanity/base/user-icon'

export default {
  name: 'person',
  title: 'Person',
  type: 'document',
  icon: UserIcon,
  fields: [
    {
      name: 'name',
      title: 'Name',
      type: 'string',
      description: 'Please use "Firstname Lastname" format',
    },
    {
      name: 'slug',
      title: 'Slug',
      type: 'slug',
      options: {
        source: 'name',
        maxLength: 100,
      },
    },
    {
      name: 'image',
      title: 'Image',
      type: 'image',
      options: {
        hotspot: true,
      },
    },
  ],
  preview: {
    select: {title: 'name', media: 'image'},
  },
}
```

The name is the object name, while the title is what will appear in the navigation within the studio. The document type is a basic building block of types within Sanity. Content types can either be document or object. A `document` type is the type you'll create and edit in the studio (i.e., they show up in the left-hand menu), whereas objects can be used to make up parts of a document (as in the `castMember` object type in this sample schema that contains a person with additional attributes that can be added to a movie). The `icon` is the icon that shows up in the left-hand navigation and is a visual cue for the type of content being created (in this case the user icon represents a person, as seen in figure 9.3).

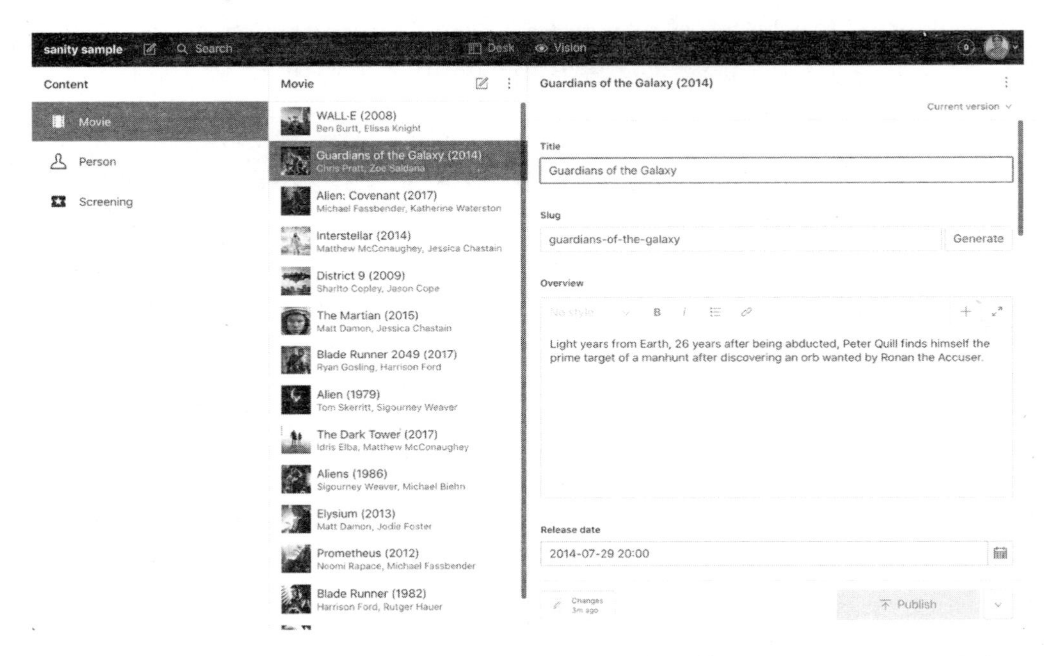

Figure 9.3 The default Sanity Studio editing experience. Sanity Studio is open source and customizable.

The `fields` are the different properties that our type contains. Each is assigned one of the 17 types. The options that can be defined in the `options` object vary depending on the type. For example, the `slug` type has a `source` option that will determine which field will be used to auto-populate a default slug value (in this case the person's `name`).

The preview section determines what fields Sanity will display in the Sanity Studio navigation when listing objects of this type. Sanity will automatically make a best guess as to which fields to display, but you can supply this to customize the fields it displays.

Feel free to continue exploring the schema to learn how it was assembled and the various options for configuring a data model in Sanity. So far, we've set up the schema and studio locally. While Sanity's system knows about it, we have not deployed it, and the studio is not available to other users. If we wanted to deploy the schema to Sanity and make the studio available, we'd run `sanity deploy` from the command line.

There are a number of example Jamstack projects that Sanity provides using SSGs, such as Next.js, Gatsby, Gridsome, Nuxt, and Eleventy. These provide examples of how to generate sites and previews using content stored in a Sanity project. All can be found at https://www.sanity.io/create.

9.3 Using WordPress

What's WordPress doing in a book about Jamstack? If you've seen the many WordPress versus Jamstack posts and discussions in recent years, you're probably thinking that you must choose one or the other. However, Jamstack doesn't have any opinions about what backend you choose, and, it turns out, WordPress offers a REST API that makes it a perfectly good backend for Jamstack sites.

It's true that the typical development scenario with WordPress involves a tightly coupled frontend and backend, but WordPress also brings a highly polished content editing experience that many content writers are already comfortable with and, in fact, may already be using. Rather than force them to move, there are a lot of options you can leverage to build a Jamstack frontend tied to a WordPress backend.

While you can integrate any Jamstack frontend with the Wordpress API, Gatsby, a very popular React-based SSG, has made it easy for developers to connect Gatsby to WordPress. Let's look at how you'd connect a Gatsby site to a WordPress instance. Don't worry if you don't have a WordPress backend set up; we'll walk through how to easily set up a local one.

9.3.1 Installing WordPress locally with Local

Let's get a WordPress instance set up first. If you already have WordPress installed, feel free to use that for this walkthrough and skip the setup (jump to "Setting Up the Gatsby Plug-Ins"). You'll need to be able to install plug-ins, which some hosts restrict.

One of the easiest ways to set up a local instance of WordPress is with a tool called Local (https://localwp.com/). While there is a paid pro version, you'll only need the free version for this walkthrough. Once installed, open Local and click the button to add a new site. You can name it whatever you'd like. I named mine "gastby-sample." Then you can choose the default "Preferred" environment. The final step is to choose a username and password for your site admin.

Local will set up the entire WordPress instance, with the MySQL database, and give you the site details. You can view the site running locally on your machine by clicking Open Site, or you can access the backend of your site running locally by clicking Admin. For now, open the admin.

9.3.2 Installing the Gatsby plug-ins for WordPress

In order to serve as a frontend to a WordPress backend, Gatsby requires two plug-ins to be installed:

- *WPGraphQL* (https://github.com/wp-graphql/wp-graphql)—The purpose of this plug-in is to take the REST API of WordPress and enable it for GraphQL (https://graphql.org/). Gatsby's data layer, which supplies the content that Gatsby uses to generate a site, is entirely based on GraphQL. Once enabled, your site will have a GraphQL API available at https://[YOUR_SITE_NAME]/graphql. For example, my site, running locally on Local, has a GraphQL API URL of http://gatsbysample.local/graphql.
- *WPGatsby* (https://wordpress.org/plugins/wp-gatsby/)—The primary purpose of this plug-in is to monitor changes on your WordPress site to keep your Gatsby frontend in sync with changes made in the WordPress backend. This plug-in monitors changes to things like posts or menus and calls a web hook in your Gatsby site to trigger a rebuild with the new information. It also enables WordPress previews to work with Gatsby pages. It is optional for our walkthrough, as it requires integration with your chosen deployment platform, like Netlify, Vercel, or Gatsby Cloud.

To install these plug-ins, open your site in Local and click on the Admin button. This will open your WordPress site's admin panel. You'll need to log in using the credentials you created earlier in the WordPress setup process using Local. Once you log in, click Plugins in the left-hand menu and then click the Add New button. In the search box, enter the names of the plug-ins, WPGraphQL (figure 9.4) and WPGatsby, and then click Install Now for each.

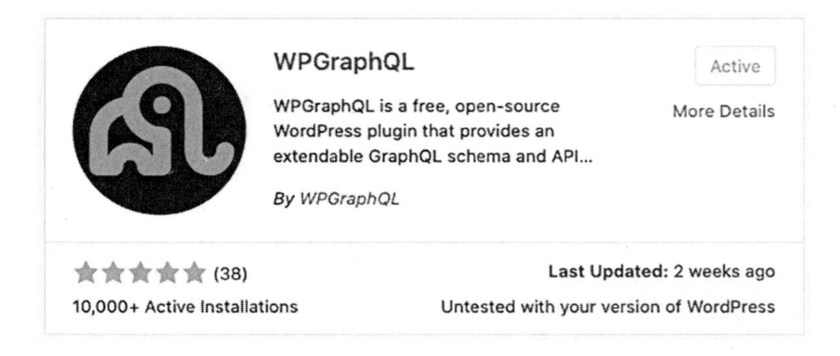

Figure 9.4 The WPGraphQL plug-in converts WordPress to work with Gatsby. This card will appear in the plug-in search UI within WordPress.

Once WPGraphQL is installed, you should see a GraphQL menu item added to the left-hand menu. Open that to view the plug-in settings, as shown in figure 9.5.

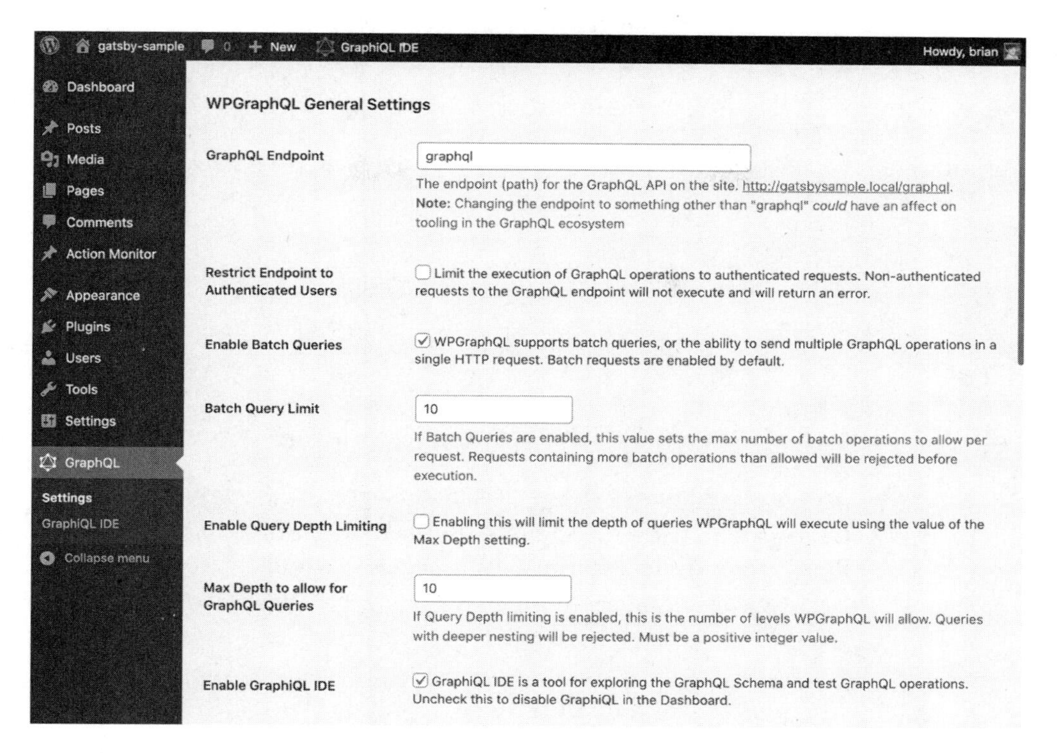

Figure 9.5 The plug-in settings screen for WPGraphQL within the WordPress admin allows you to set the endpoint location and change how your GraphQL endpoint will work.

Make note of the GraphQL endpoint URL linked beneath the GraphQL endpoint setting. This is the endpoint our Gatsby frontend will connect to. For example, mine is http://gatsbysample.local/graphql. While you are testing, you can also scroll down and check the Enable GraphQL Debug Mode box. This will enable debug information to be returned by GraphQL queries during development.

By default, WPGraphQL also added a GraphiQL IDE query editor to your WordPress admin, which is available via the GraphiQL IDE menu item on the top navigation bar. Click that to open a query editor and test GraphQL queries against your WordPress data.

The full schema that you can query against will load in the Explorer on the left. It's pretty complex, but the Explorer lets you click around and create queries. Feel free to review this more on your own, but for now let's test it with a query similar to what we'll use in our simple example Gatsby app.

Listing 9.2 A query to get WordPress content from WPGraphQL

```
query MyQuery {
  posts {
    edges {
```

```
        node {
          title
          author {
            node {
              name
            }
          }
          slug
          excerpt
          content
        }
      }
    }
  }
```

Click the play button (i.e., the right arrow) above the query editor to run the query.

Listing 9.3 The results returned by the query in listing 9.2

```
{
  "data": {
    "posts": {
      "edges": [
        {
          "node": {
            "title": "Hello world!",
            "author": {
              "node": {
                "name": "brian"
              }
            },
            "slug": "hello-world",
            "excerpt": "<p>Welcome to WordPress. This is your first post.
              ➥ Edit or delete it, then start writing!</p>\n",
            "content": "\n<p>Welcome to WordPress. This is your first post.
              ➥ Edit or delete it, then start writing!</p>\n"
          }
        }
      ]
    }
  },
  "extensions": {
    "debug": []
  }
}
```

Figure 9.6 shows how this looks in the GraphiQL IDE.

That's all we need to do to set up WordPress. Feel free to populate the site with some posts or simply proceed using the default posts that WordPress provides.

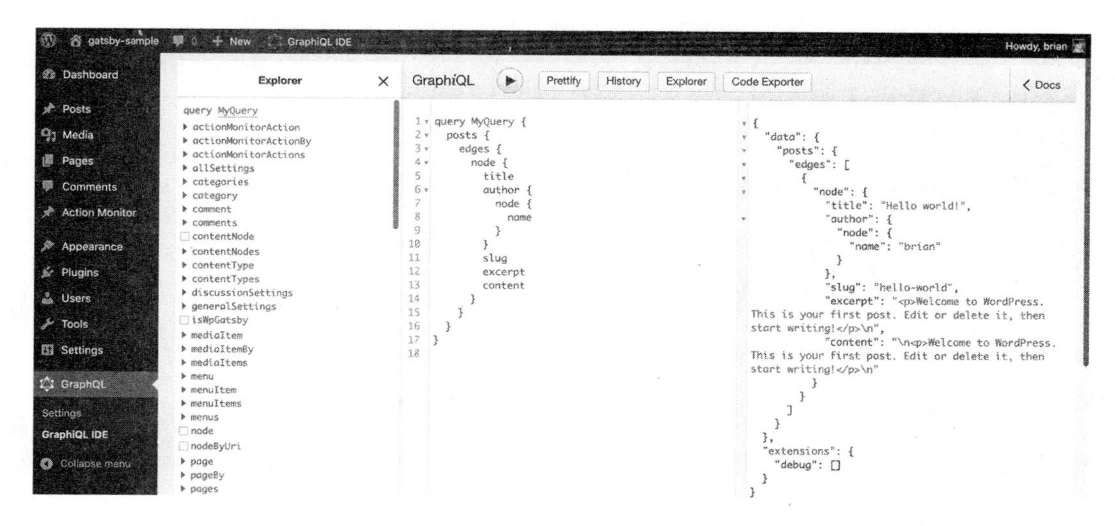

Figure 9.6 Querying the GraphQL API generated by WPGraphQL within the GraphiQL interface, inside the WordPress admin

9.3.3 Setting up Gatsby

There are a number of Gatsby starters that are designed to work with WordPress and will jumpstart your project. You can find those by going to the Gatsby Starters page (https://www.gatsbyjs.com/starters) and filtering by "CMS: WordPress." However, for the sake of better understanding how all this works, we're going to write our own simple Gatsby landing page that will list the posts within our local WordPress site.

Before we can connect Gatsby to WordPress, we'll need to initialize a new Gatsby project using the CLI. To do this, open your project directory and run the following command (we'll specify the folder we want the Gatsby project created in during the setup process):

```
npm init gatsby
```

This will run `create-gatsby`, which will ask you a series of questions via the command line to configure your new Gatsby site. Here are the responses you'll need to create our sample project, which are illustrated in figure 9.7:

- *What would you like to call your site?* Gatsby WordPress.
- *What would you like to name the folder where your site will be created?* gatsby-wordpress.
- *Will you be using a CMS?* Use your arrow keys to choose WordPress.
- *Would you like to install a styling system?* No (or I'll add it later).
- *Would you like to install additional features with other plugins?* Use the arrow keys to scroll down and choose Done.

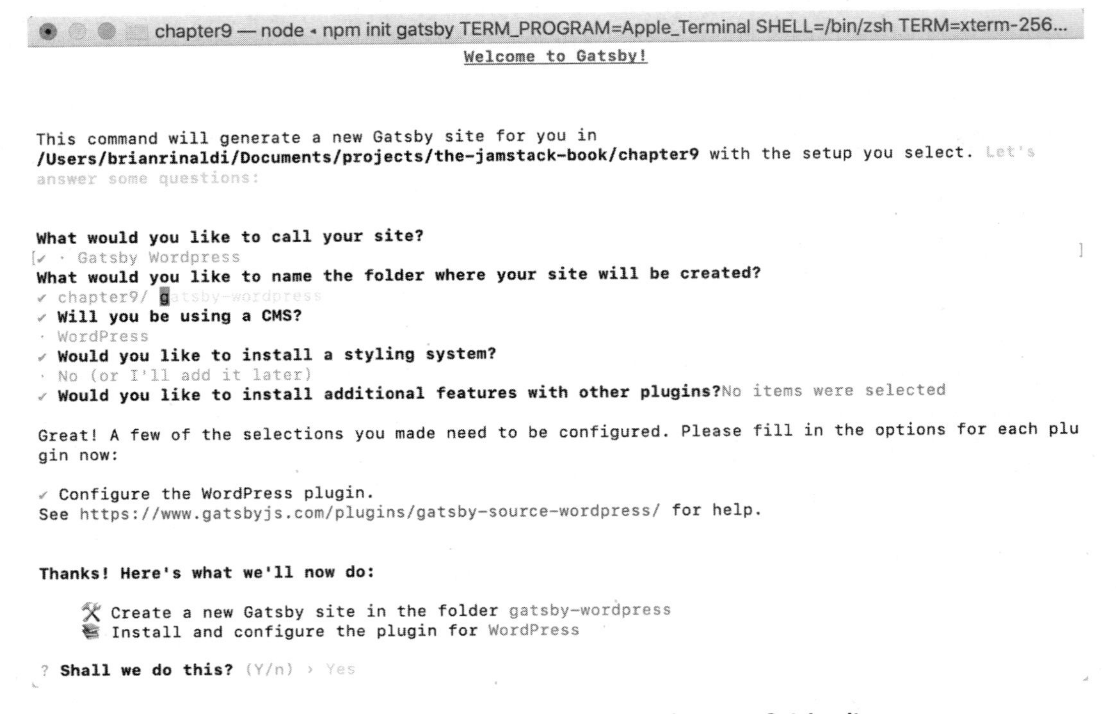

Figure 9.7 The Gatsby CLI provides a step-by-step setup process for a new Gatsby site.

- *Configure the WordPress plugin.* This is the GraphQL endpoint we noted from the GraphQL settings within our local WordPress instance. For example, mine is http://gatsbysample.local/graphql.
- *Shall we do this?* Press Enter to select yes.

This will start the process of generating the default files for our Gatsby site in the gatsby-wordpress directory. Once this is done, change the directory to the project folder and start Gatsby. Note that, due to an issue with the process in creating our Gatsby app, we may need to install the `gatsby-plugin-sharp` and `gatsby-transformer-sharp` plug-ins before running Gatsby:

```
cd gatsby-wordpress
npm install gatsby-plugin-sharp gatsby-transformer-sharp
npm run develop
```

Once the build is done, we can view our site at http://localhost:8000/, which should look like figure 9.8.

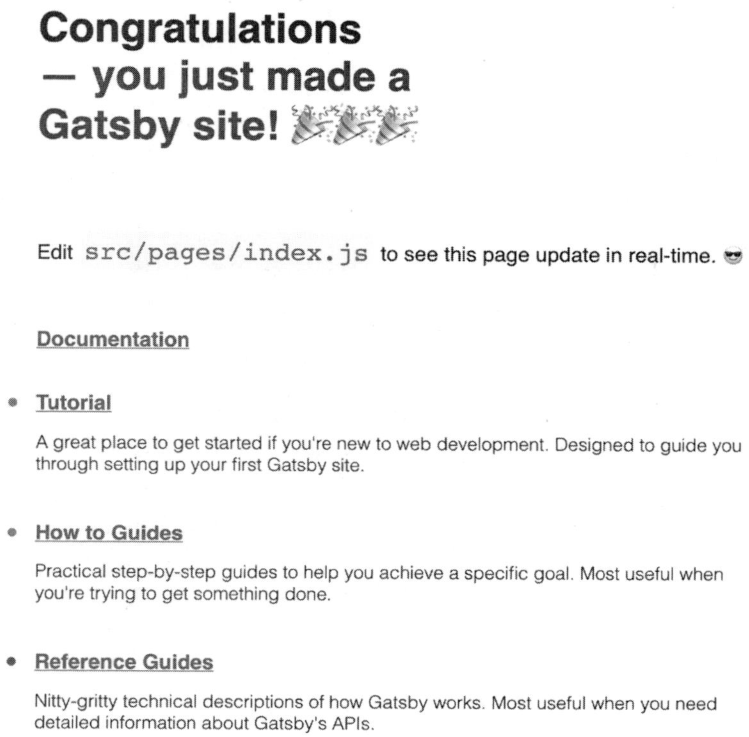

Congratulations — you just made a Gatsby site! 🎉🎉🎉

Edit `src/pages/index.js` to see this page update in real-time. 😎

Documentation

- Tutorial

  A great place to get started if you're new to web development. Designed to guide you through setting up your first Gatsby site.

- How to Guides

  Practical step-by-step guides to help you achieve a specific goal. Most useful when you're trying to get something done.

- Reference Guides

  Nitty-gritty technical descriptions of how Gatsby works. Most useful when you need detailed information about Gatsby's APIs.

Figure 9.8 The default home page of the site generated by Gatsby's CLI

9.3.4 *Exploring Gatsby's data layer*

Before we begin coding, let's explore Gatsby's data layer. Gatsby prefers to get all data for your application from a GraphQL-based data layer. We can view the data and queries available in this data layer when our site is running locally by going to http://localhost:8000/graphql. Let's open that in a new browser window.

Since we already explored the GraphiQL editor that WPGraphQL provided, this should look somewhat familiar. In this case, the queries in the Explorer panel include all of Gatsby's default data queries as well as a number of WordPress queries. The latter were added to Gatsby's data layer because we chose WordPress as our CMS during the site creation process.

Feel free to explore the queries available, but for now, let's test the query we'll use to populate data on our home page. Enter the query in the query editor and click the Play button to run it.

Listing 9.4 A query to retrieve WordPress content from within Gatsby's data layer

```
{
  allWpPost {
    edges {
```

```
        node {
          title
          author {
            node {
              name
            }
          }
          slug
          excerpt
          content
        }
      }
    }
  }
```

This query is very similar to the one we ran inside the WordPress GraphiQL editor, except that it uses the `allWpPost` query that it is provided by Gatsby's data layer.

Listing 9.5 The results returned from the query in listing 9.4

```
{
  "data": {
    "allWpPost": {
      "edges": [
        {
          "node": {
            "title": "Hello world!",
            "author": {
              "node": {
                "name": "brian"
              }
            },
            "slug": "hello-world",
            "excerpt": "<p>Welcome to WordPress. This is your first post.
              ➥ Edit or delete it, then start writing!</p>\n",
            "content": "\n<p>Welcome to WordPress. This is your first post.
              ➥ Edit or delete it, then start writing!</p>\n"
          }
        }
      ]
    }
  },
  "extensions": {}
}
```

Perfect! We're ready use this data in our home page.

9.3.5 Consuming WordPress content in Gatsby

Open the project files in your code editor. The root project folder contains a gatsby-config.js that was generated for us during the creation process. It contains the name of the site that we specified as well as the WordPress plug-in configuration that we provided with our GraphQL endpoint URL. If we were to deploy our WordPress instance, this is where we'd update the endpoint URL:

```
module.exports = {
  siteMetadata: {
    title: "Gatsby Wordpress",
  },
  plugins: [
    {
      resolve: "gatsby-source-wordpress",
      options: {
        url: "http://gatsbysample.local/graphql",
      },
    },
  ],
};
```

The source of our site is under the src directory. Within that directory, the pages for our site are within the pages directory. We only have a home page (index.js) and a 404 page (404.js). Let's open index.js and make some changes.

The page uses data that is defined inline as a `links` array and is right below the styles. We won't need that, so let's remove it. Let's also remove the code to render the links list from the markup so that our `IndexPage` markup looks like the following:

```
// markup
const IndexPage = () => {
  return (
    <main style={pageStyles}>
      <title>Home Page</title>
      <h1 style={headingStyles}>My Blog</h1>
      <ul style={listStyles}></ul>
    </main>
  );
};
```

At this point, our page isn't rendering anything but a `h1` with the text "My Blog." Before we can render the list of posts, we need to make the data available to the page. Let's create the query that will be run against Gatsby's GraphQL data layer directly beneath the `export default IndexPage` line. This query will look familiar, as it is the same one we tested earlier when exploring Gatsby's data layer.

First, we need to add an import at the top of our index.js file:

```
import { graphql } from 'gatsby';
```

Then we can add the query:

```
export const pageQuery = graphql`
  query IndexQuery {
    allWpPost {
      edges {
```

```
          node {
            title
            author {
              node {
                name
              }
            }
            slug
            excerpt
            content
          }
        }
      }
    }
  `;
```

Next, let's make the data from this query available to the page. First, we need to add another import at the top of the index.js file:

```
import PropTypes from 'prop-types';
```

Then we can tell our Gatsby page about the data we're providing via our `allWpPost` query. The `PropTypes` library is type checking to ensure the data matches what we expect. Place the following code after the `export default IndexPage` line:

```
IndexPage.propTypes = {
  data: PropTypes.shape({
    allWpPost: PropTypes.shape({
      edges: PropTypes.array,
    }),
  }),
};
```

Now we can update our markup to output the results from the query. The code destructures data from our page props and then uses the array of items within `data.allWpPost`, which contains the results of our query, to output our posts as a list with excerpts. Note that we have to use `dangerouslySetInnerHTML` to display the excerpt content because it is returned as HTML. This property is named this way because doing so makes DOM changes outside the scope of React's virtual DOM, which is fine for our purposes but should generally be done judiciously.

Listing 9.6 Outputting a list of WordPress blog posts within Gatsby

```
// markup
const IndexPage = ({ data }) => {
  return (
    <main style={pageStyles}>
      <title>Home Page</title>
      <h1 style={headingStyles}>My Blog</h1>
      <ul style={listStyles}>
```

```
        {data.allWpPost.edges.map((post) => (
          <li key={post.node.slug}>
            <span>
              <a style={linkStyle} href={post.node.slug}>
                {post.node.title}
              </a>
              <p
                dangerouslySetInnerHTML={{ __html: post.node.excerpt }}
                style={descriptionStyle}
              ></p>
            </span>
          </li>
        ))}
      </ul>
    </main>
  );
};
```

Our updated home page running in the browser should look like figure 9.9, with the one post from our default WordPress install displayed, along with an excerpt from the post.

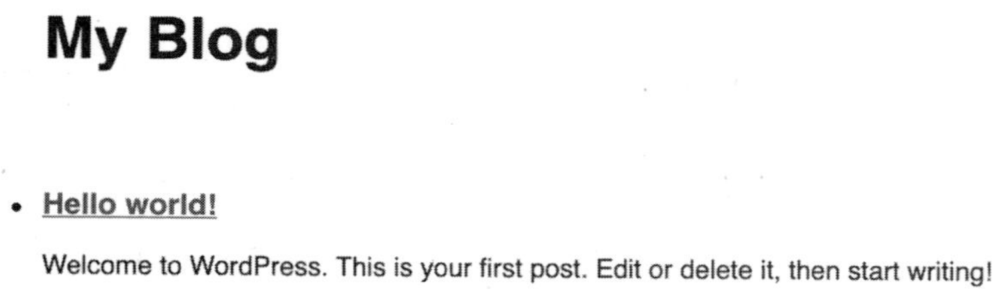

My Blog

- Hello world!

Welcome to WordPress. This is your first post. Edit or delete it, then start writing!

Figure 9.9 The edited blog home page outputs the blog posts and descriptions within our WordPress instance. By default, WordPress installs with a single dummy post. Add more posts to see them appear.

9.3.6 *Next steps in using WordPress as a headless CMS*

We're using WordPress as a CMS to populate our Gatsby site. The next step in building this application is to create the dynamic route to handle displaying the individual WordPress posts in Gatsby. While we won't cover that here, I encourage you to explore more about how to integrate Gatsby and WordPress via Gatsby's WordPress guide (https://www.gatsbyjs.com/guides/wordpress/), or this tutorial on CSS-Tricks (http://mng.bz/4jEv) covering the Gatsby WordPress source plug-in.

It is important to remember that the WordPress REST API can be consumed by any Jamstack frontend, whether Gatsby or not. In fact, the GraphQL endpoint provided by the WPGraphQL plug-in can be consumed outside of Gatsby as well, using either Eleventy or Next.js, for example. While Gatsby provides some built-in connections for WordPress, WordPress is also a valid headless CMS backend for just about any Jamstack site, regardless of what SSG you choose.

9.4 Connecting a CMS using a site builder

Up to now, we've talked about ways you can connect a headless CMS using some of the starters that are provided by either the SSG or the headless CMS provider itself. We also looked at how to connect your own using WordPress. However, there is one more route I want to explore, which is using a site builder like WeWeb (https://www.weweb.io/), Strattic (https://www.strattic.com/), or Stackbit (https://www.stackbit.com/).

Site builders like these go a few steps beyond what a starter template provides by customizing the site that it generates based on your needs and providing web-based tools for either technical or nontechnical resources to edit that site. They also allow you to integrate the content management via popular existing headless CMS providers.

Keep in mind that these services are generally free to start but do charge you once you reach usage caps or want to lift feature restrictions.

9.4.1 WeWeb

WeWeb is a Vue-based site generator, so it does not rely on an SSG, that can generate a site that is fully editable directly in the browser. This means that you aren't simply editing text content but can also add and rearrange elements on the page from within its web-based WYSIWYG editor (figure 9.10).

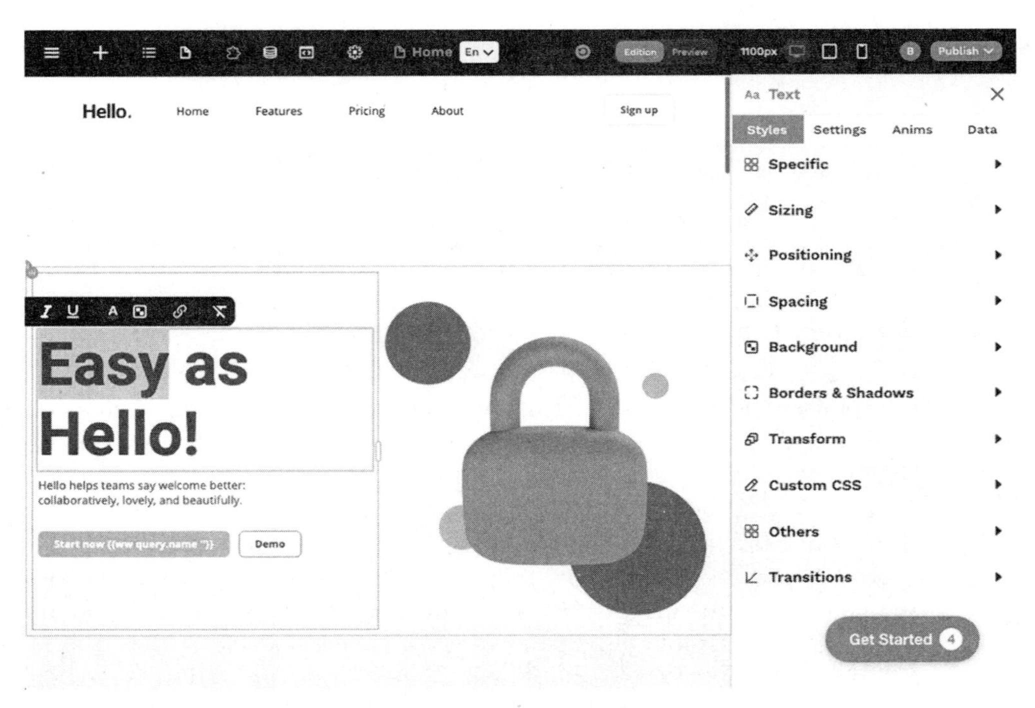

Figure 9.10 The WeWeb admin UI allows you to edit the text on the page and the properties that affect the look and feel of the web design.

You can build and customize a site based on prebuilt templates that WeWeb provides or start from a blank site. In both cases, you'll be able to utilize the preexisting site components to create and modify your site. These components allow for dropping in common website elements like contact forms or rich text. You can also create and upload your own custom WeWeb components to use.

WeWeb has built-in connections to a number of headless CMS offerings, including Strapi, Ghost, or WordPress. Other data sources it provides are Airtable, any REST API, Google Sheets, any RSS Feed, any GraphQL endpoint, an SQL database, or any endpoint you can connect to using JavaScript.

9.4.2 Strattic

Strattic takes a very different approach to building a Jamstack site connected to a CMS. Essentially, Strattic deploys a full WordPress setup for you on its servers. You manage content as you normally would via the WordPress admin. Strattic even supports most standard WordPress themes and many plug-ins, though due to the nature of the static site files, you'll want to check compatibility (http://mng.bz/QW6Q) with your favorite plug-ins first.

When you are ready to publish, you use the installed Strattic plug-in to deploy a Jamstack version of the site to Strattic's servers (figure 9.11). The site looks and feels just like a WordPress site, but it is built with static files and serverless APIs. These serverless APIs make things like site search, comments, forums, and other dynamic functionality work.

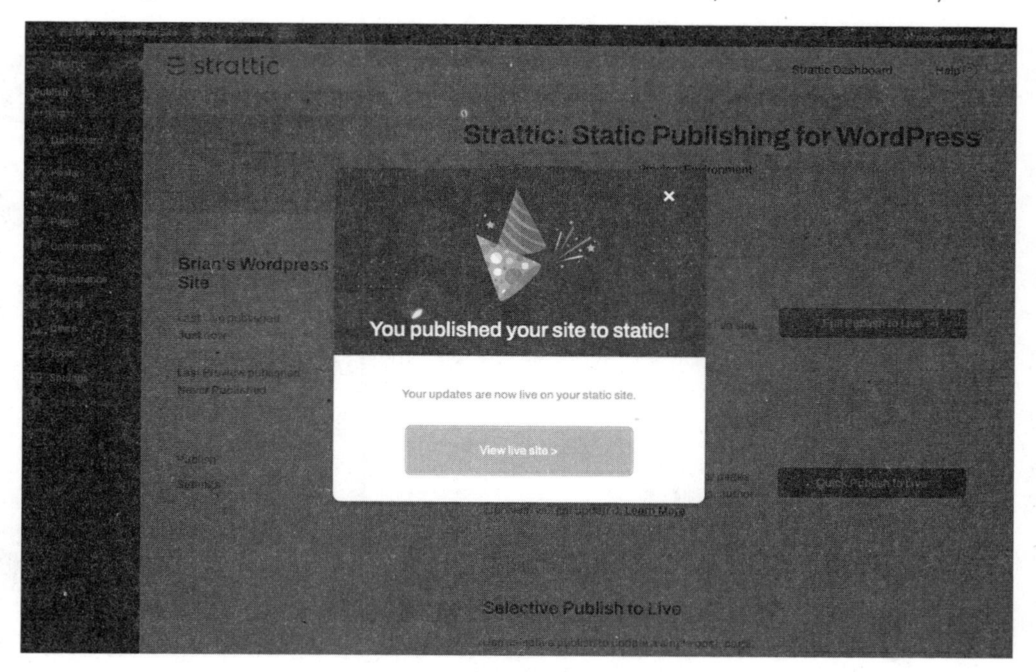

Figure 9.11 The Strattic WordPress plug-in provides several publishing options for your site. Once the static site code is generated and deployed, you'll be able to view it running on Strattic's servers.

Keep in mind that even though this is a Jamstack site, by most people's definition, you don't have access to the frontend files. However, Strattic does give you SFTP access to the WordPress PHP files. Strattic offers a free trial but requires a paid account after that expires.

9.4.3 Stackbit

Stackbit follows a more traditional Jamstack approach than WeWeb in that it leverages SSGs within its site builders. In fact, it supports multiple SSGs as well as multiple headless CMS options. The site builder is free to use, and the code is pushed to your own GitHub account and deployed on your Netlify account. However, the optional site editing suite does require a paid account to use a number of the features.

The first step in building a Jamstack site using Stackbit is to choose a theme (figure 9.12). There are 17 different themes to choose from as of this writing, based on a number of archetypes, such as a blog or e-commerce. You can add a custom theme, but this does require the proper stackbit.yaml configuration file that tells Stackbit how to map the content on your page to the content in a CMS.

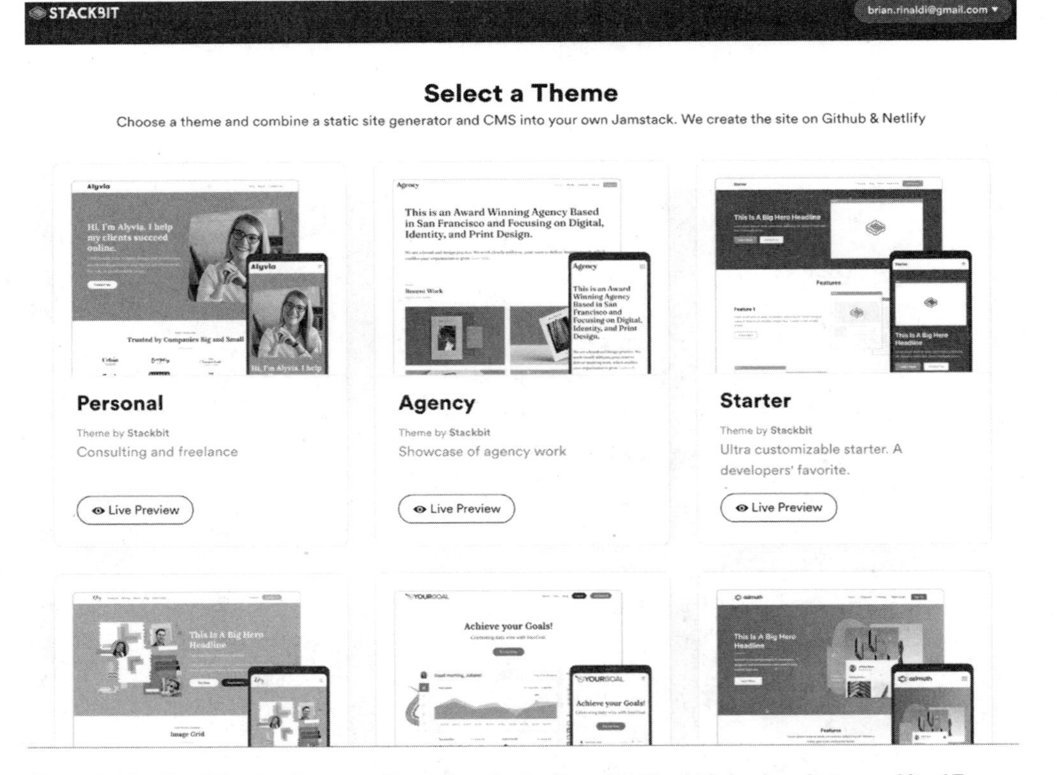

Figure 9.12 The first step in generating a Jamstack site using Stackbit is choosing one of its 17 available themes or uploading a custom theme of your own.

Next, you are brought to a summary screen that allows you to modify the other default site build options, as seen in figure 9.13. The default SSG is Next.js; however, you can also choose Jekyll, Hugo, or Gatsby. It's worth noting that not all the available headless CMSs are currently compatible with Next.js within Stackbit's system.

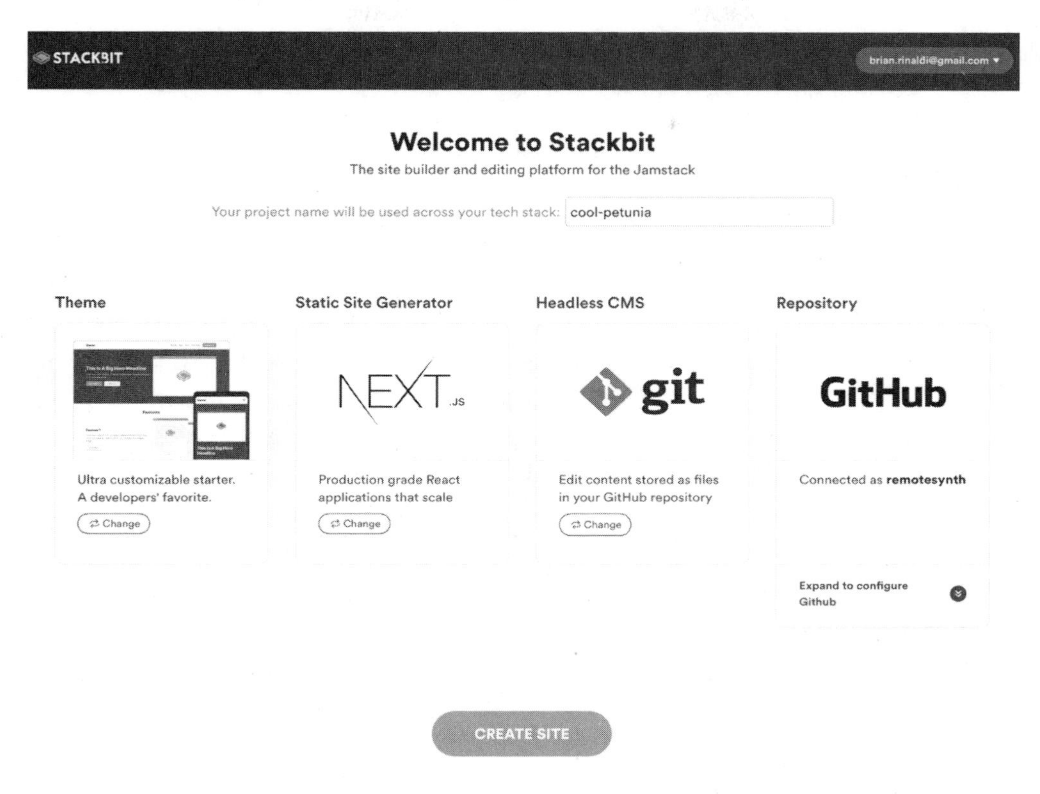

Figure 9.13 Once you've chosen a theme, you can customize the SSG and headless CMS that Stackbit will use. All the code will be pushed to your connected GitHub account.

The default CMS is Git. This is a pure Git-based CMS that manages content directly in your connected Git repository. It handles draft and preview changes within Stackbit's editor by maintaining a separate branch for unpublished changes. While this option is completely free, you'll want to consider whether a Git-based CMS is the right fit for your project, and you'll likely require a paid Stackbit account if you have multiple content contributors/editors.

The other headless available options are Sanity, Contentful, Netlify CMS, Dato CMS (an API-based CMS), and Forestry (a Git-based CMS).

You'll need to connect Stackbit to your GitHub account in order to create a site, but, once that is complete, clicking the Create Site button will generate all the site files using your chosen theme and SSG, as well as provision your headless CMS and populate some default content. If you've connected your Netlify account, it will also deploy on Netlify. When the site is done generating and deploying, you'll be brought into Stackbit's content editor. The editor is free to use for one user, with some feature limitations, but it's worth noting that the editor is entirely optional. You can modify your code directly in GitHub, and your content directly in your chosen CMS.

9.5 What's next?

We've explored a few different options available to you when choosing a headless CMS, and even looked at some tools and resources for integrating these. As we discussed in the beginning of the chapter, there are about 87 different options for headless CMS, so it may feel overwhelming to make an informed choice. Here are just a handful of questions to ask yourself to help refine the options worth evaluating:

- *Does the file-based editing of a Git-based headless CMS fit my project, or do I need the added flexibility of an API-based headless CMS?* We discussed the pros and cons of each earlier in the chapter to help you decide which option best suits your project.
- *Does the solution need to be open source?* If so, a solution like Netlify CMS or Strapi is worth investigating. You can find other open source options by going go Jamstack's list of headless CMSs and filtering by open source licenses.
- *Do we have an existing CMS that I can leverage as headless (e.g., WordPress)? If so, do we want to continue to maintain and manage the CMS ourselves or consider moving to a hosted third-party service?* In many cases, when it is possible, keeping your existing CMS will be the most straightforward solution, as it doesn't require retraining all of the content contributors on a new system. This can also ease getting buy-in on moving to the Jamstack from other stakeholders.
- *Does the editing experience meet the needs of my content editors? How difficult will it be for them to learn?* If a user hasn't worked in Markdown before, it can be a difficult transition. If a user is used to page-based editing (like in WordPress), then a move to a modular system used by an API-based CMS can take some time and training. It may be that these difficult transitions for content contributors are unavoidable, but you should at least be aware of them and plan for them.
- *Does the CMS offer integrations with my SSG that will reduce the difficulty of connecting our Jamstack frontend to the headless CMS?* Not every CMS has templates for every SSG, but many do. Be sure to investigate whether a prebuilt integration already exists before you start building one from scratch.

Obviously, there will be more questions specific to the needs of your project and your team, but even just asking these five will, hopefully, help you narrow down the options before evaluating them.

For a good list of additional considerations when choosing a headless CMS, check out "How to Choose a Headless CMS" by Emmanuel Tissera in *Smashing Magazine* (http://mng.bz/aD57).

Summary

- There are two main types of headless CMSs: Git-based and API-based. Git-based CMSs store content in files that are managed within your Git repository and edited via an admin UI. API-based CMSs store and edit the content within their systems and provide it to your application via an API.
- The primary benefit of a Git-based headless CMS is cost. The primary benefit of an API-based headless CMS is the ease of content reuse.
- Contentful is a popular API-based headless CMS that is configured and managed entirely via its admin UI. Sanity is another popular API-based headless CMS, but it is configured entirely via JavaScript and edited via the open source Sanity Studio.
- WordPress provides an API that can be used as the backend for a Jamstack site. Gatsby provides a number of tools for integrating the API into the Gatsby data layer that provides all the data and content for generating sites using Gatsby. We walked through getting a local WordPress site connected to a new Gatsby site using these tools.
- Site builders offer an easy, though generally not free, way to generate a Jamstack site that can connect to a variety of headless CMSs. We explored three site builders: WeWeb, Strattic, and Stackbit.

10 Migrating to the Jamstack

This chapter covers

- Understanding migration challenges
- Solving build-time issues on large sites
- Strategies to migrate existing site functionality
- Exploring popular third-party Jamstack services
- Tips and strategies from the Jamstack community for making a migration

You've made it to the last chapter, and hopefully, by now you're convinced about the benefits of using the Jamstack for your web development projects. You may even want to migrate an existing project to the Jamstack. The good news is that this is entirely feasible and, in fact, quite common. The bad news is that there's no one-size-fits-all tool or wizard that can handle the migration; it is, for the most part, a customized solution for every migration.

There are so many factors that go into a migration, each one unique. This chapter aims to give you some guidance for planning your migration, as well as some resources and tools that can help ease the process. This process is more than just a technical challenge. It can involve a lot of stakeholders and other organizational considerations that you'll have to manage, but we will primarily focus on the technical migration aspects.

10.1 What kind of site are you migrating?

Often developers want to just jump in, pick a tool, and start building—especially when they are excited about utilizing new technologies—but it's critical to first take a step back and evaluate your existing site. This will guide you in choosing the right tools for the job. Jamstack is a fast-moving area of development, with lots of new tools that generate a lot of developer interest. But just because a particular solution has a lot of buzz doesn't mean it is the right tool to build your application. However, just because a Jamstack tool has been around for a while doesn't disqualify it as a solution, either.

Let's look at three broad categories of sites that your project may fall into:

- Content-heavy sites
- Web applications
- Large sites

Keep in mind that your project may fall into more than one category. For instance, you may have large, content-heavy sites, or different parts of your site may fall into unique categories. For example, you may have a content-heavy information site that is also a web application accessible after the user authenticates.

10.1.1 Content-heavy sites

Many, if not most, websites are focused on delivering content rather than functionality. This category could include everything from a blog, product, or service documentation; a large corporate site with a lot of informational pages; or a news site that constantly updates. Here are some things to consider when migrating a content-heavy site:

- *Do I need to migrate existing content?* This depends on what you are migrating from and what you need to migrate to. For instance, in chapter 9 we looked at how you can integrate the headless WordPress API into an existing Jamstack site. Drupal, another popular CMS, can also be used as a headless CMS (http://mng.bz/XWyp). In fact, many monolithic CMSs can be used as headless backends, so be sure to see whether that is an option for you.
- *Are there tools to help with the migration?* If you are migrating to an API-based headless CMS, some offer migration tools for popular monolithic CMSs. Be sure to explore the viability of that option first. If you are migrating to file-based content that can be stored in a Git repo and managed by a Git-based headless CMS, there are tools that can help with that as well, though many are community-maintained. For example, Hugo maintains a list of community-built migration tools (https://gohugo.io/tools/migrations/), including from sources like Joomla, Drupal, and WordPress.

- *Does our current CMS implementation include a lot of customization?* Despite these tools, it is possible (if not likely) that you'll need to write a custom migration script in order to handle all the edge cases. In my experience, years of CMS use often build up a legacy of "shortcodes," plug-ins, and other highly customized usages of the CMS that can require extra care during a migration. Take stock of any plug-ins and shortcodes that may be used to modify your content before writing your migration script. A common example is that WordPress automatically renders a YouTube video player from a YouTube URL. If there are a lot of plug-ins and shortcodes but the existing CMS has an API providing the rendered HTML, it may be easier to transition the rendered content rather than the raw content.

10.1.2 Web applications

A web application is defined not by its content, though it will likely have content considerations, as well, but instead by its functionality. Here are three things to consider:

1. *Do you need to migrate existing functionality, or can you leverage existing backend APIs?* If your backend code is already accessible via APIs that you can call, then you may not need to transition it (at least not immediately anyway). If you need to migrate, you'll need to decide whether to do so via an existing third-party Jamstack service or write custom code (e.g., as serverless functions deployed to the cloud and called from the application). We'll deal with these questions more in section 10.2.
2. *Do you need a frontend JavaScript framework? What requirements are there for your frontend? Will the application benefit from building on one of the SSGs that are built on frameworks like React, Vue, and Angular?* The answer for many web applications may be yes, but not always. There are still a lot of web applications that can save a lot of overhead on the client by using plain JavaScript instead of a frontend framework. Frontend frameworks can do some heavy lifting in terms of loading data on the client and managing the state on the application, making it easier to build many UI interactions, but they come with overhead in terms of the amount of JavaScript your visitors are required to download.
3. *What can you prerender?* Keep in mind that one of the key things that differentiates the Jamstack approach is prerendering wherever possible. When it comes to web applications, it can be easy to default to rendering everything at run time by making server/API calls. Some creative thinking can help illuminate portions of the site that may not be as dynamic as they seem. New techniques, such as distributed persistent rendering (DPR) or incremental static regeneration (ISR) (which we'll talk more about in section 10.1.3), can also help address things like user-generated content. Keep in mind that a page does not need to be fully prerendered (during the build) or fully client-side rendered. You can render parts of the page during the build, and other parts can load on the client via JavaScript.

10.1.3 Large sites

Large sites have one unique problem when it comes to the Jamstack: build times. This may seem trivial, but if you are building a site with thousands or even hundreds of thousands of pages (as some e-commerce sites have), build times can get especially lengthy. Even a build time that is only a few minutes can become a real impediment to developing and testing a site. They can also interfere with content updates. Imagine the reaction of a marketing team waiting on a 15-minute build to push a small but critical correction to the site, for example.

Sean C. Davis looked at this issue in an article for CSS Tricks in late 2020 (so it doesn't account for some recent updates to the SSGs) titled "Comparing Static Site Generator Build Times" (http://mng.bz/y4Dy). You can see the performance he found in his tests for large sites in figure 10.1.

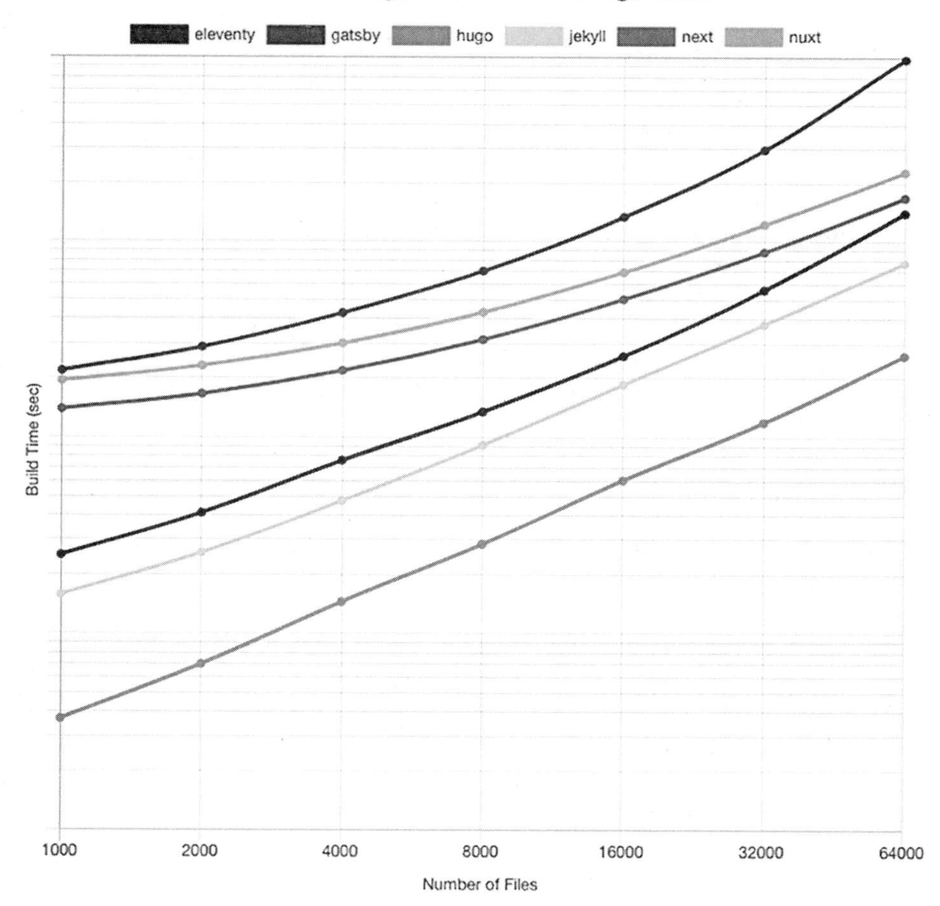

Figure 10.1 This chart shows how each SSG scales in build times for large sites. Some, like Hugo, have a linear increase, while others increase exponentially as the number of pages increases.

As figure 10.1 shows, the more "traditional" SSGs—those not based on a JavaScript framework and that focus entirely on generating static files—tend to have faster build times that increase linearly. The JavaScript-framework-based SSGs tend to take longer to build and increase exponentially as sites get particularly large.

Thus, the simplest solution to the issue of build time performance is to choose a SSG that is optimized for fast build times, such as Hugo or Eleventy (although Jekyll also performed surprisingly well in Davis's tests). Hugo is built in Go and is designed for extremely fast build times—even for sites with tens of thousands of pages. Eleventy is built in JavaScript and, while not as fast as Hugo, it is still extremely fast for large sites and includes some performance-related tooling to help debug long-running build processes.

Both Hugo and Eleventy are "traditional" SSGs, though. This means that they focus on generating static pages, don't require a framework, and don't run on both the client and the server (in the case of Jamstack, the "server" is a serverless function). If you cannot choose one of these options, there are solutions built into some of the other SSGs that can help.

INCREMENTAL BUILDS

Gatsby supports *incremental builds* (http://mng.bz/M2Ko). In simple terms, this means that Gatsby tracks the impact of any changes and only rebuilds the pages that are impacted rather than regenerating the entire site on every rebuild. Gatsby says that its tests show this reduces a 20-minute build time to under a minute for the first build and under 10 seconds for subsequent builds. Support for incremental builds in Gatsby, though, depends on your deployment provider. For instance, Gatsby's own Gatsby Cloud obviously supports incremental builds, as does Netlify (http://mng.bz/aDzo). It is worth noting that Jekyll also has experimental support for incremental builds (see http://mng.bz/g4Dx).

INCREMENTAL STATIC REGENERATION (ISR)

ISR is an optimization supported by Next.js (see http://mng.bz/5KEZ for a detailed description). The easiest way to understand how it works is to compare it to incremental builds. Gatsby's incremental builds happen during the build after each deploy by regenerating only updated pages, meaning that at some point all the pages need to be generated before they can be tracked and, if necessary, updated. However, using ISR, you'd only generate a subset of pages that you initially specify and update missing pages *after* they are deployed, when they are first requested by the user.

Essentially, ISR is a way of deferring the generation of noncritical pages to when a user requests them. The requested pages are added to the build, and each subsequent request will be served the static version, meaning only the initial visitor will see any delay in receiving the content. This can occur in one of two ways:

- *With a fallback*—The user requests a page that has not yet been generated. They will receive a static page in a loading state (i.e., any content and data has not been populated) and the page will update once the server responds with the generated page. All subsequent requests will receive the completed page via the static file, not the loading state.
- *Without a fallback*—This is the recommended alternative. The user requests a page that has not yet been generated. The server will respond to the user when the page is rendered. Subsequent requests will serve the static file.

One important distinction of ISR is that you can set a time to revalidate. This means that Next.js will show the stale version but regenerate the page in the background. When the page has generated, it will invalidate the cache and show the updated page to the current visitor. All subsequent visitors will see the updated page served from the static file until the time to revalidate has passed again. While this is similar to server-side rendering (SSR), the key difference is that an SSR page will hit the server on every request, while an ISR page will only hit the server when a page is stale and will serve the static file otherwise. Functionally, it is very similar to SSR plus a cache.

ISR will allow a large site to generate only the critical pages, reducing the number of pages that need to be built initially, and thereby reducing the build time. For instance, an e-commerce site with thousands of products could generate the most frequently trafficked pages and defer rendering for the others in order to drastically reduce their overall build time. Currently, ISR for Next.js is only supported when hosted on Vercel (http://mng.bz/0wxz), although Netlify supports a similar variation called distributed persistent rendering, which we'll discuss next.

DISTRIBUTED PERSISTENT RENDERING

Netlify has always supported what they call atomic and immutable deploys.

- *Atomic deploys*—A deploy is made as a single, complete unit with no interim state. This means that the prior deploy exists until the new deploy is complete and nothing gets updated in between. Thus, if a deploy fails, the site doesn't go down, as the prior deploy was never replaced in whole or in part.
- *Immutable deploys*—Once a deploy is complete, it cannot be changed after the fact. This allows you to roll back to a prior deploy, for instance, without anything breaking.

As Netlify saw it, ISR breaks these contracts, which they consider a critical benefit of Jamstack deployment (though there is disagreement in the community as to how critical they are to the Jamstack). To address this, Netlify proposed a standard they call *distributed persistent rendering* (DPR) (http://mng.bz/6ZrZ).

The key differences between ISR and DPR are as follows:

- *DPR was proposed as a standard.* (You can view the RFC here: http://mng.bz/KBYn.) The idea is that, in doing so, it can be adopted and supported both by multiple SSGs and deployment platforms.
- *DPR does not support an invalid/stale state for a page.* Deferred pages are rendered once upon first request and added to the build, thereby becoming a persistent asset of the build. This means that rolling back to a prior build would include any deferred pages that had been previously rendered as they become part of the deploy. It also means there is no question as to the state of the rendered page; it is either rendered or not yet rendered but never out of date. Netlify believes this avoids hard-to-debug issues related to the state of a build when using ISR.

Ultimately, DPR solves the same problem as ISR by allowing large sites to defer rendering pages and decrease their build time. Currently Netlify supports DPR for Next.js sites via their Essential Next.js build plug-in (http://mng.bz/9K28), which installs automatically for Next.js deployments. As of this writing, only Eleventy has publicly shared that it is actively working to support DPR as well via Eleventy Serverless (http://mng.bz/jyG9).

DECIPHERING THE TERMINOLOGY

I just threw a lot of terminology at you: incremental builds, ISR, and DPR. It can be tough to know how to solve the problem of long builds for large sites. Let's break it down:

- If you are building a large site, explore the possibility of using either Hugo or Eleventy. While neither includes a JavaScript framework by default and neither currently supports incremental builds, ISR, or DPR, both are optimized for building large sites. If you've determined that your project requires a JavaScript framework, then it is worth exploring other options (but remember that even though they haven't been built on the backend using a framework, you are still free to include a framework on the frontend with either solution).
- If you are using Gatsby, be sure to choose a deployment provider that supports incremental builds.
- If you are using Next.js, your choice of ISR or DPR is largely dictated by your deployment platform (e.g., Vercel or Netlify). They are both functionally similar and will achieve similar results in reducing build times, but DPR will not allow you to specify a time to revalidate. This means that, once generated, a page cannot be updated when using DPR.
- If you are using an SSG other than Hugo, Eleventy, Gatsby, or Next.js, your options for reducing build times on large sites are limited at the moment. Most of the techniques described here are new, and other popular SSGs are working on adding support, so this may change in the near future.

10.2 What functionality do you need to migrate?

Jamstack sites rely on JavaScript and API calls to enhance the frontend with dynamic functionality, which promotes the use of discreet services to accomplish these tasks. In some cases, this is backend code you write yourself as cloud (i.e., serverless) functions. These functions may contain custom business logic that your site requires, or they may make API or database calls and then transform the data for consumption by the frontend. These custom functions can be deployed as Netlify functions, Cloudflare functions, AWS Lambda, and more. The choice generally depends on your chosen stack and your chosen deployment provider.

But migrating to a Jamstack site does not necessarily require that you convert all your existing server-side code into custom cloud functions. Jamstack generally encourages the use of purpose-built third-party services that are designed to meet many common site requirements like authentication, search, data storage, and more. In many cases, these can save you significant time and effort when building a project.

In order to determine the best available options, start by breaking down the functionality of your existing site into services. Then consider the following questions:

- *Can we leverage existing backend code via an API?* Moving to the Jamstack doesn't mean rewriting everything from scratch. If some of your existing backend code already exposes an API or can be easily modified to expose an API that can be consumed in your Jamstack application, this may be the best available option because requires the least development effort and allows other stakeholders to continue utilizing existing systems. Nonetheless, you might consider transitioning to another option anyway if the Jamstack migration presents an opportunity to refactor and improve legacy code.
- *Can I use a third-party service?* Even if you may not be able to leverage existing backend code, you don't need to build your backend from scratch. There are a plethora of services that can be easily used in Jamstack sites. The main considerations should be whether they meet your requirements in terms of features and pricing. Many services have free developer accounts that scale up in cost based on usage, so it's important to consider pricing before making a change. See section 10.2.1 for a description of a number of popular options.
- *Does the third-party service offer migration tools?* If you are choosing a third-party service, be sure to explore whether they offer a migration path from your existing backend. This is especially common when migrating content from popular monolithic CMSs, like WordPress or Drupal, to a headless CMS. However, other services may offer migration tools depending on what you are migrating from (e.g., authentication providers often offer migration paths), so be sure to do some research before doing the work yourself.
- *How do I break down custom code into discrete services?* The way existing server-side code in your monolithic web application can translate to serverless cloud functions may not be immediately obvious. Break down the functionality of your site

and take an inventory of what needs to be written. In particular, look for opportunities to create reusable pieces of code. Try to keep the purpose of each cloud function simple, reusable, and discrete.

10.2.1 Popular third-party Jamstack services

There are far more options available for nearly every category of service than I can cover here, but it is worth exploring some of the available options to give you a better sense of the sort of third-party solutions that may be available to you.

- **Content management**
  - *Contentful*—An enterprise-focused, API-based headless CMS that emphasizes ease of content modeling and a polished editing experience.
  - *Sanity*—An API-based headless CMS that focuses on the flexibility and customizability in both its content modeling and editing experience.
  - *Agility CMS*—An enterprise-focused, API-based headless CMS that offers a large number of prebuilt integrations with popular SSGs.
- **E-commerce**
  - *Snipcart*—An add-on headless e-commerce solution that is designed to be extremely easy to integrate into any Jamstack site with minimal effort.
  - *Commerce.js*—An API-based headless e-commerce solution that is designed for creating highly customized e-commerce experiences.
- **Authentication/user management**
  - *Auth0*—An all-in-one authentication and user management solution that integrates with most existing third-party login services as well as custom options.
  - *Netlify Identity*—A solution that can work with external providers and is designed to integrate quickly and easily within any site deployed to Netlify.
- **Forms**
  - *Netlify Forms*—Captures the contents of any form submission and designed to be easy to add to any custom form on sites deployed to Netlify.
  - *Formspree*—A flexible form option that can integrate easily into added to HTML forms or integrated via React components or the API.
- **Search**
  - *Algolia*—A hosted search index and API that provides prebuilt integrations for most popular frontends (and backends).
  - *Lunr*—An open source client-side Jamstack search library built into JavaScript that can quickly perform searches against a generated index file.
- **Media management/processing**
  - *Cloudinary*—Provides media storage and APIs that can modify, transform, and optimize images and video before displaying them on your site.

> **Looking for more options or categories?**
> If you're looking for additional options in these categories or for services in categories that I didn't cover here, there are a couple of really useful resources. The folks at CloudCannon, a commercial Git-based headless CMS, have created a Jamstack Ecosystem site (http://mng.bz/W7IX) filled with services in a wide array of categories, including feature comparison charts. It's an easy way to get a quick sense of the options available. While it doesn't cover as many categories, Discovery (https://bejamas.io/discovery/), a site by the folks at the Bejamas consultancy, goes into great detail about headless CMS, SSG, and hosting options.

10.3 Making the move

You've audited your site to figure out the different pages, templates, and functionality that you'll have to accommodate. You've explored SSGs and headless CMS options, and you've researched the supporting services that you may want for things like authentication/user management, search, forms, and more. Finally, you've sold your boss and any stakeholders on the value of this project, and now they are looking to you for next steps. While I can't give you a project plan, as it depends on far too many factors for me to accurately cover here, I can give you a few more recommendations and guidance with a little help from some Jamstack community experts.

10.3.1 Don't move everything at once

> *Start small to learn and experiment first. List all the features you currently have and see how it would translate in a Jamstack world. It's not just about picking a SSG, sometimes you'll need to employ many services to achieve a similar result. Evaluate the feasibility and cost.*
>
> —Frank Taillandier (personal communication)

> *Most sites are bigger than you think, with lots of hidden or weird features. Also, most older sites are monolithic. Don't try to recreate EVERYTHING at once—you can do things in a phased approach as you split up frontend, backend, content, APIs, etc. Use something Netlify or Vercel rewrites to combine old pages or URL paths into a single domain.*
>
> —Joel Varty (personal communication)

This is critical: you can adopt the Jamstack incrementally. The entire site doesn't have to be converted and moved to the Jamstack all at the same time. If you're dealing with a particularly large site, break it down into pieces that can be implemented individually. You can leverage redirects and proxies, which most deployment providers, such as Netlify and Vercel, support, so that the transition is seamless for your users; they don't need to know which parts of the site exist on the provider's CDN and which exist on your legacy servers. Or you could host the Jamstack portions of your site on your existing infrastructure since it is just static HTML, JavaScript, and CSS files; there's no additional infrastructure required to support it.

In addition, you can begin transitioning services and deploy cloud functions that can be used on your existing site and your new site simultaneously. This may require a bit more work up front by adding two different site integrations but can save you headaches in the long run. For example, transition some server-side logic into an AWS Lambda function that can be used both by the new Jamstack sections deployed and by the preexisting portions of the site. Starting small can have the added benefit of giving you time to experiment and learn, particularly if you are new to the Jamstack, without overburdening yourself.

10.3.2 Pick a headless CMS up front

> *Don't forget about your marketing and editing teams who will need to use your solution for ages and iterations—make sure they can create and edit pages, that they have modules, that they can manage URLs and sitemaps, that they have a reliable preview and functional text editor, that they can search and find content easily, that they can reuse modules, that they can create forms and landing pages—you don't build sites just for 100 speed score to show off on your Twitter. You build them for humans who need to feel efficient, creative, and independent.*
>
> —Olga Voigt (personal communication)

You may be tempted to simply manage your content as files in Markdown. Markdown is easy. Git already versions files for you. Why bother with the added complication and expense of moving to a headless CMS?

If you have multiple content contributors/editors or separate teams working on site code versus site content, this will, in my opinion, ultimately be a mistake that reveals itself over time. Even worse, it becomes more difficult to fix as time goes on because the volume of content that needs to be transitioned to a CMS only grows.

In my experience, the problem occurs because, unless you are very careful, the processes for deploying content and code become mixed when deploying content as files. Code changes get mixed in with content changes, causing problems on both ends. Developers can't push their code because of content changes that aren't ready to go live, while content editors are often blocked from getting critical content out because of code changes that are not ready to be released.

While an API-based headless CMS obviously solves this, even a Git-based headless CMS allows for a publishing process that doesn't require mixing processes. For instance, Netlify CMS supports a simple workflow process for content approvals, allowing you to push content changes directly into production once they have been reviewed and approved. This keeps the workflow for content separate from the regular deployment flow for code changes, allowing each to work independently.

10.3.3 Consider building templates from scratch rather than porting

> *If you're coming from WordPress, I would export the content and build from scratch or start with a Jamstack template. Scraping a WordPress theme and converting it to an SSG isn't going to be a fun time.*
>
> —Mike Neumegen (personal communication)

It may be tempting to think you can port or scrape your existing templates into the templating system supported by your SSG, but this can be painful and not take advantage of the frontend flexibility that Jamstack offers. Instead, you may want to consider building new templates from scratch (perhaps using the opportunity for a design refresh) or use one of the countless themes available for nearly all SSGs.

For example, if you were using WordPress, you might consider porting each template from PHP or even scraping the generated HTML and converting that to a template. The idea is to save yourself a lot of time and effort, but the porting process can be difficult and time-consuming. For instance, some templating solutions supported by popular SSGs, like Liquid or Handlebars, don't include logic that translates easily from a language like PHP. Even in cases where the templates do support that kind of logic, for example JavaScript/JSX, translating them can be extremely time consuming, negating the benefit of doing it in the first place.

Plus, when you port a template, you are less likely to properly take advantage of converting reusable portions of your templates into components or partials. Or you may not take full advantage of the componentized content offered by your headless CMS when you are converting templates designed for page-based content. For example, it is not uncommon to find elements of design and layout within the content for a page stored in WordPress. When moving to a headless CMS, it is best practice to separate the design and layout elements from the content by building this page using multiple content elements rather than a single block of content.

> **Access hundreds of free, open source themes**
> For a comprehensive list of themes that covers a variety of SSGs and headless CMSs, check out JamstackThemes.dev. All of the themes are open source and free. They offer an excellent starting point for a new or existing project.

10.3.4 Keep as much as possible

I would advocate for keeping as much as possible. Like a lot of people might still want to use WordPress. Good for them, if that's what they're comfortable with. Use the WPGraphQL plug-in and consume the GraphQL output through Gatsby or combine with another service with an API mesh.

—Jaden Baptista (personal communication)

The decoupled nature of Jamstack means that you can, and in many cases should, keep parts of your existing backend in place. There is no requirement that you move everything to a new Jamstack service or move everything upfront. There may be parts of your backend that either meet a very specific need and will be difficult to port or that already support being used in a headless manner and are easy to port.

For example, you may be using WordPress, Drupal, or another monolithic CMS. If your team is happy with the content management and editing experience, it is worth considering using your existing CMS in a headless manner. On the other hand, your site may integrate data from a specialized backend that stores data in a database, CRM (customer relationship management) system (e.g., Salesforce), or in an ERP (enterprise resource planning) system (e.g., SAP). While this would be complex to port into another system, explore whether you are able to use existing APIs or, alternatively, connect the database to services that can expose it (e.g., services like Hasura [hasura.io], StepZen [stepzen.com], or TakeShape [takeshape.io] can expose existing databases as GraphQL APIs that your Jamstack frontend can consume).

10.4 What's next?

Well, we've reached the end of our Jamstack journey together, but hopefully just the beginning of yours. If I were to leave you with one final tip, it's that there is a great community that has been created around the Jamstack and is always willing to help, so you don't have to go on that journey alone. Start by joining one (or both) of the existing online communities for Jamstack:

- *Jamstack Community Discord* (https://jamstack.org/discord)—This is a large Discord community that is managed by Netlify but open to everyone; there's no requirement that you use Netlify.
- *The New Dynamic Slack* (visit https://www.tnd.dev/ and click the Slack link in the upper-right corner)—This is a community-managed Slack group that has been around for some time. While it is much smaller, its active members include many of the creators of popular Jamstack tools.

Another opportunity to meet and learn from fellow Jamstack developers is via meetups and conferences around the globe. You can search your local area on Meetup.com or browse the list of user groups on https://jamstack.org/community/. There are also a couple of Jamstack-specific conferences, including the Jamstack Conference (https://jamstackconf.com/), organized by Netlify, and TheJam.dev (https://cfe.dev/events/the-jam-2022/), organized by one of our coauthors, Brian, in *Jam*uary (get it?).

The Jamstack community is, in my experience, one of the strengths of Jamstack in general. Be sure to participate in it, and I will see you there.

Summary

- No two migrations are alike, but there are some common issues that content-heavy sites, web applications, and large sites face when moving to the Jamstack. These include migrating content, migrating functionality, and dealing with long build times.
- Content-heavy sites should decide whether they need to migrate their existing content and, if so, the best headless CMS option to migrate to. Some headless CMS systems have migration tools for moving from popular monolithic CMS solutions. Determine whether your existing CMS supports a headless deploy-

ment. If so, determine whether your needs are better met staying with the existing solution or moving to a dedicated headless CMS solution.

- Web applications must take an inventory of what functionality needs to be transitioned and determine what existing backend code can be reused in the Jamstack site, what code needs to be rewritten, and what services may be available to replace existing functionality. This inventory can be used to explore whether you can reuse any existing backend APIs, whether you should migrate to a third-party solution, or whether you need to build your own backend.
- When moving a large site to the Jamstack, developers need to take build times into consideration. The available options for dealing with this are choosing a highly optimized SSG like Hugo or Eleventy, or choosing a solution that supports ISR or DPR, such as Next.js.
- There is a large and helpful community of developers around Jamstack who can be a great resource when you are learning to build Jamstack sites or when you hit a roadblock.

index

A

B

C

G

H

I

J

L

M

N

O

P

R

S

RELATED MANNING TITLES

Hugo in Action
Static sites and dynamic Jamstack apps
by Atishay Jain
Foreword by Steve Francia

ISBN 9781617297007
488 pages, $59.99
Spring 2022

Svelte and Sapper in Action
by Mark Volkmann

ISBN 9781617297946
456 pages, $59.99
September 2020

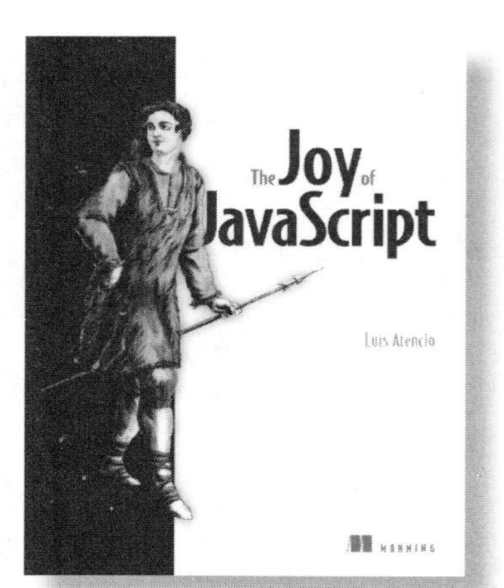

The Joy of JavaScript
by Luis Atencio

ISBN 9781617295867
360 pages, $39.99
February 2021

For ordering information, go to www.manning.com

The Manning Early Access Program

Don't wait to start learning! In MEAP, the Manning Early Access Program, you can read books as they're being created and long before they're available in stores.

Here's how MEAP works.

- **Start now.** Buy a MEAP and you'll get all available chapters in PDF, ePub, Kindle, and liveBook formats.
- **Regular updates.** New chapters are released as soon as they're written. We'll let you know when fresh content is available.
- **Finish faster.** MEAP customers are the first to get final versions of all books! Pre-order the print book, and it'll ship as soon as it's off the press.
- **Contribute to the process.** The feedback you share with authors makes the end product better.
- **No risk.** You get a full refund or exchange if we ever have to cancel a MEAP.

Explore dozens of titles in MEAP at www.manning.com.